Dr Catherine Dawson has been a researcher for almost thirty years, working in the public, private and higher education sectors. She has a Master's degree in social research and a doctorate in adult education, and has obtained funding for a variety of research projects during her career. These include topics such as public perception of change in higher education, the learning choices of adult returners, basic skills education for prisoners, student housing in the South East and students and alcohol misuse.

Dr Dawson has designed and delivered research methods courses at university and run research methods training courses for workers in the public, private and charitable sectors. This includes courses for individuals and organizations wishing to obtain funding for their research. She has also written extensively for academic journals and is the author of *Introduction to Research Methods* and *Advanced Research Methods*, which are aimed at beginning and experienced researchers.

Also published by Constable & Robinson

Introduction to Research Methods

Basic Study Skills

Advanced Research Methods

Writing a UCAS Personal Statement

Writing Your Dissertation

How To Write an Assignment

How To Finance Your Research Project

Dr Catherine Dawson

A How To Book

ROBINSON

ROBINSON

First published in Great Britain in 2015 by Robinson

ISBN 978-1-84528-571-5 (paperback)
ISBN: 978-1-84528-572-2 (ebook)

Typeset by SX Composing DTP, Rayleigh, Essex
Printed and bound by CPI Group (UK) Ltd, Croydon, CR0 4YY

Robinson
is an imprint of
Constable & Robinson Ltd
100 Victoria Embankment
London EC4Y 0DY

An Hachette UK Company
www.hachette.co.uk

www.constablerobinson.com

How To Books are published by Constable & Robinson, a part of Little, Brown Book
Group. We welcome proposals from authors who have first-hand experience of their
subjects. Please set out the aims of your book, its target market and its suggested
contents in an email to Nikki.Read@howtobooks.co.uk

Contents

PART 2 COSTING PROJECTS

PART 3 MAKING APPLICATIONS

PART 4 ACTING ETHICALLY

Preface

Research funding can be very difficult to obtain, especially for researchers who do not work in high-profile, research-intensive universities. Competition is fierce, with researchers pitted against each other, often having to compete with friends and colleagues for available funds. This can result in a large number of good projects remaining unfunded.

However, globally there is a vast amount of money available for research. Through careful searching, selection, preparation and application, it is possible to be successful in winning grants for your research. This book will help to ease the grant application process by providing detailed information about sources of funding, choosing appropriate funding organizations, producing budgets and making a successful application. It includes practical tips, exercises and checklists that will help you to apply for, and succeed in winning, your research grant.

The book is aimed at the following types of researcher:

- academic researchers working in universities and research centres;
- student researchers who need to fund their research;
- early-career researchers seeking more information about the funding process;

- researchers working in industry and in research and development departments;

- contract research organizations;

- government and political researchers;

- library, museum and archive researchers;

- health and medical researchers;

- researchers working for charitable and not-for-profit groups;

- researchers working for community groups;

- self-employed and retired researchers;

- research managers.

All academic disciplines and subject areas are covered, including the sciences, social sciences, humanities, medicine, health and education. The information provided is relevant to researchers working in the United Kingdom, United States, European Union and further afield, and covers government-funded, private-funded and charity-funded research.

The book is divided into four parts that guide you through the funding process. Part 1 offers advice about finding funds. It discusses the different types of funds that are available and is aimed at specific types of researcher, such as those working within education or industry. This section concludes by highlighting the importance of choosing the right funding organization and ensuring that your research is relevant and has a chance of being funded.

Part 2 goes on to offer advice about costing a project. This can include large-scale research that requires a very detailed breakdown of direct and indirect costs, and small-scale projects needing only a simple budget. Specific advice is offered about calculating staff, estate and equipment costs and knowing how to work out costs for collaborative projects. The importance of budget justification is highlighted.

Part 3 deals with making a grant application. Some grant applications will need to follow a specific format, whereas other funders are more flexible with the application structure and style. However, all researchers will need to highlight the importance, innovative nature and impact of their research. This section also describes the submission process and offers advice about what to do if proposals are rejected.

Part 4 offers advice about acting ethically. This includes information about acting in accordance with researcher codes of ethics, avoiding conflict of interest, avoiding biased financial relationships with funders and submitting proposals to research ethics committees. Regrettably, some funding organizations have specific agendas that can compromise the research process: this section includes guidance about what you can do to prevent this happening.

The book concludes with comprehensive lists of funding databases and directories that will help with your funding search, and a list of further reading and resources for those who wish to follow up any particular issues in more detail.

PART 1
FINDING FUNDS

Chapter 1

Knowing About Sources of Funding

There are various funding options available for researchers who wish to obtain a grant for their research project. Money from public sources is available from organizations such as research councils, the National Health Service (UK), the National Institutes of Health (US and CAN), various government departments and the Technology Strategy Board (UK).

Private sources include charities, trusts, industry and wealthy individuals. It is also possible to obtain finance from internal sources such as university departments and research and development units within private industry.

FUNDING FROM PUBLIC SOURCES

Public (or state) funding is provided by governments to invest in science and research. The way that the funding is organized and administered depends on the country. In the UK, for example, the budget for science and research funding is allocated by the Department for Business, Innovation and Skills (BIS) and is organized via the Dual Support System into two main channels:

- the seven research councils that provide grants for specific projects and programmes (see over);

- the higher education funding bodies that provide block grant funding to universities.

The Haldane principle

Public funding for science and research in the UK is said to operate under the 'Haldane principle'. The original concept was that the decision about what to spend research funds on should rest with the researcher, rather than with politicians. However, there have been several revisions to this idea since its inception (it is thought to have been derived from Lord Haldane's 1918 report on the machinery of Government). Today, the UK Government 'identifies strategic priorities, and the scientific community selects projects within relevant fields on the basis of scientific merit, as assessed by peer review'. A statement of the principle, published by the Government, can be viewed at www.publications.parliament.uk.

Some of the major public funding bodies from the UK and overseas are described below. For information about specific grants that are available for different types of researcher, see Chapters 3, 4, 5 and 6.

THE ROYAL SOCIETY:
PUBLIC FUNDING SHOULDN'T BE RING-FENCED

In 2013 Davis Willets, the universities and science minister, announced that eight areas of research and development would be targeted to stimulate economic growth. These areas included robotics and autonomous systems, advanced materials, grid-scale energy storage and practical applications for graphene. However, The Royal Society believes that this type of ring-fencing can lead to low-quality research and stifle innovation, encouraging researchers to follow funding streams rather than inspiration and innovation. This is part of a long-running debate about public funding for research: who makes decisions; how are their decisions made; and are they the best people to make those decisions?

The research councils (UK)

The UK's seven research councils invest around £3 billion across all academic disciplines, including medical and biological sciences, astronomy, physics, chemistry and engineering, social sciences, economics, environmental sciences and the arts and humanities. Funding for space programmes is allocated by the UK Space Agency, which became an executive agency of BIS in 2011. A list of all the research councils and their websites can be found in Chapter 2.

Research councils provide funds to all eligible UK higher education institutions, research institutes and independent research organizations. See Appendix 1 for a list of all the organizations that can receive funding from UK research councils.

The European Research Council (EU)

The European Research Council (ERC) is an independent body set up in 2007 to fund investigator-driven research in the European Union (EU). The total budget allocated to the ERC for the period 2007 to 2013 was €7.5 billion under the European Union's Seventh Framework Programme for Research and Technological Development (FP7). Between 2007 and 2013, the UK received €4.9 billion from this programme, which represented around 9 per cent of the total funding available.

The next ERC calls will be made under the future programme (Horizon 2020), which has taken over from FP7 for the period 2014 to 2020. This has a budget of nearly €80 billion spread across three pillars: excellent science, societal challenges and industrial leadership. The excellent science strand encompasses grants and fellowships from the ERC (see Chapter 3 for more information about this programme).

The main goal of the ERC is to encourage high-quality research in Europe through competitive funding. The ERC will fund researchers of any nationality and any age and supports 'proposals that cross disciplinary boundaries, pioneering ideas that address new and emerging fields and applications that introduce unconventional, innovative approaches'. More information about ERC funding can be obtained from erc.europa.eu/funding-and-grants.

The Australian Research Council (AUS)

The Australian Research Council (ARC) is a statutory agency that manages the National Competitive Grants Program. The ARC supports fundamental and applied research and research training through national competition across all disciplines, with the exception of clinical medicine and dentistry. It also brokers partnerships between researchers and industry, government, community organizations and the international community.

Funding rules are produced each year and include information about the scheme, eligibility requirements, the application process, selection procedures, approval processes and requirements for the administration of funding. For more information visit www.arc.gov.au.

The National Academies (UK and US)

In the UK BIS is responsible for government funding of important programmes at three of the UK's National Academies: the Royal Academy of Engineering (www.raeng.org.uk), the British Academy (www.britac.ac.uk) and The Royal Society (royalsociety.org). Examples of grants from these academies are provided in Chapter 3.

In the US there are four National Academies: the National Academy of Sciences, the National Academy of Engineering, the Institute of Medicine and the National Research Council. Most research is carried out by the National Research Council, with other academies providing 'high-quality, objective advice on science, engineering and health matters'. Funding can be from federal sponsors, government agencies, non-governmental organizations or non-profits, for example. More information can be obtained from www.nas.edu.

The National Institute for Health Research (UK)

The UK Government funds health-related research through two main strands: the National Institute for Health Research (NIHR) and the Medical Research Council (MRC: see Research Councils, above). The NIHR is a government body, established in 2006, that commissions and funds National Health Service (NHS), social care and public health research. The key objective is to 'improve the quality, relevance, and focus of research in the NHS and social care by distributing funds in a transparent way after open competition and peer review'.

Chapter 3 provides examples of the type of research funding that is available through these bodies. More information can be obtained from www.nihr.ac.uk and www.mrc.ac.uk respectively.

The National Institutes of Health (US and CAN)

The National Institutes of Health (NIH) funds medical research in the US and around the world. It is part of the US Department of Health and Human Services and is made up of twenty-seven institutes and centres (a list of these can be found on the NIH website: www.nih.gov). Each year around $31 billion is invested in medical research by NIH, awarded mainly through competitive grants and to scientists working on projects in NIH laboratories. Information regarding funding opportunities can be found on the grants page at www.grants.nih.gov.

In Canada the Canadian Institutes of Health Research (CIHR) is the federal funding agency for health research. It is composed of thirteen institutes, listed on the CIHR website: www.cihr-irsc. gc.ca. An up-to-date list of funding opportunities is available from ResearchNet: www.researchnet-recherchenet.ca.

Government departments (worldwide)

Government departments throughout the world fund specific types of research. In the UK, for example, the Department for International Development (DFID) funds research that can lead to new technologies and better ways of helping the poorest people in the world. This includes research into new drugs for malaria and sleeping sickness, better diagnostic tests for tuberculosis and vaccines for diseases in cattle in Africa.

Visit Research4Development (R4D) (www.r4d.dfid.gov.uk) to find out more. This is a free-to-use online portal containing the latest information about research funded by DFID, including details of current and past research in over 35,000 project and document records.

The Technology Strategy Board (UK)

Public funding for research, development and commercialization in the UK is organized by the Technology Strategy Board (TSB), an executive non-departmental public body reporting to BIS. It offers a range of funding programmes and works with businesses, universities and other organizations. The main funding schemes for research available from the TSB are:

- **catalyst** (a form of research and development funding that focuses on a specific priority area and aims to help take projects from research to commercial viability);

- **collaborative R&D** (encouraging businesses and researchers to work together on innovative projects in strategically important areas of science, engineering and technology);

- **feasibility studies** (a grant scheme that allows businesses the opportunity to test an innovative idea and its feasibility to be developed and taken to market);

- **innovation vouchers** (encouraging businesses to look outside their current network for new knowledge that can help them to grow and develop);

- **smart** (offering co-funding to UK-based pre-start-ups, start-ups, micro businesses and small and medium-sized enterprises [SMEs], to carry out science, engineering and technology R&D projects that could lead to successful new products, processes and services);

- **the small business research initiative** (providing 100 per cent funding through a contract to small businesses in the UK).

Chapter 4 provides detailed information about each of these funding schemes, and more information about all the TSB schemes can be obtained from www.innovateuk.org.

If you are a private sector or industry researcher in South Africa, similar programmes are offered through the Technology and Human Resources for Industry Programme (THRIP) (thrip.nrf.ac.za). If you work for a small or medium-sized enterprise in Canada, funding can be obtained from the Industrial Research Assistance Program (www.nrc-cnrc.gc.ca).

FUNDING FROM PRIVATE SOURCES

Private sources of funding for research include charities, trusts, industry and wealthy individuals. Rules, regulations, terms and conditions vary, depending on the funding organization, type of grant and the country in which you live. You will need to obtain guidelines and eligibility criteria from chosen organizations to ensure that they are suitable before you consider making an application (see Chapter 7).

Trusts and charities play an important role in research across the world. Although, in some cases, they might not fund the full economic cost of research (see Chapter 8), they do provide an independent source of funding that enables researchers to carry out their work and generate new knowledge. These organizations are also keen to plough that knowledge back into further research that will continue to be of benefit to society.

Chapter 2 provides information about finding sources of private funding and directs you to some of the major funding organizations in specific countries and worldwide. Chapters 3, 4, 5 and 6 provide information about specific grants that are available. In addition you can use the databases and funding directories listed in Appendices 2 and 3 for more details about sources of private funding.

THE WELLCOME TRUST: FUNDS FOR PUBLIC ENGAGEMENT

In 2013, the Wellcome Trust announced its intention to dedicate up to £4.5 million a year to public engagement. Researchers can apply for funds, which have no upper or lower limit, while applying for mainstream scientific grants. Researchers will need to submit high-quality proposals that specify who they want to talk to, why they wish to talk to them and the methods that they intend to use. The initiative has been established to improve and enhance public engagement with scientific research.

For more information, visit the Wellcome Trust website: www.wellcome.ac.uk.

FUNDING FROM INTERNAL SOURCES

As we have seen above, in many countries state funding is allocated to universities and national academies as block grants that are then distributed by those bodies in the form of scholarships, fellowships or other types of specialist research grant. Institutions advertise and administer the funds so you will need to apply direct to the institution. Contact your chosen university research office for more information.

Other sources of internal funding may be available, such as funds originating from company profit, philanthropy, public donations or fees to use a service. Contact your industry research and development unit, research office or line manager to find out more.

SUMMARY

Various sources of funding are available for research, including public sources distributed through research councils and block grants, and private sources through trusts, charities, industry and wealthy individuals. Other funding is available through internal sources such as university departments and research and development units.

Searching for public and private funds can be a long and laborious process. However, the next chapter offers advice about carrying out an efficient and effective funding search, and directs you to some useful public and private funding body websites.

Chapter 2

Carrying Out a Funding Search

There are many funding bodies throughout the world that offer both small and large grants to researchers. However, finding out what is available can be a lengthy and difficult undertaking, especially for researchers who are new to this process.

This chapter offers practical advice about how to find information, including using databases and directories, seeking advice from professionals and through networking. It also provides a useful list of some of the larger public and private funding bodies and their websites.

USING DATABASES AND DIRECTORIES

Databases and directories provide a quick and easy way to search for funding. Most will enable you to perform a keyword search that can be narrowed down by subject area, category or agency, for example. As a researcher seeking funding, the following databases may be of use (see Appendix 2 for more details):

- Charity Choice (UK) www.charitychoice.co.uk

- Euraxess Funding Search (UK) euraxessfunds.britishcouncil.org

- Foundation Center (US) foundationcenter.org

- Funding Central (UK) www.fundingcentral.org.uk

- Funding Finder (UK) www.biglotteryfund.org.uk/funding/funding-finder

- Grants.gov (US) www.grants.gov

- Grant Tracker (UK) www.grant-tracker.org

- Pivot (worldwide) pivot.cos.com

- Research Professional (UK) www.researchprofessional.com

- The Directory of Social Change (UK) www.trustfunding.org.uk, www.governmentfunding.org.uk, www.companygiving.org.uk, www.grantsforindividuals.org.uk

- The Australian Competitive Grants Register (AUS) education.gov.au

- The Australian Directory of Philanthropy (AUS) www.philanthropy.org.au

- The Directory of Development Organizations (worldwide) www.devdir.org

- The International Cancer Research Partnership (ICRP) database (worldwide) www.icrpartnership.org/database.cfm

- The National Science Federation (US) www.nsf.gov

- The Scholarship Portal (EU) www.scholarshipportal.eu

- The Wheel (IRE) www.wheel.ie

Printed directories also contain comprehensive listings of funding organizations (most are updated annually). A list of these directories can be found in Appendix 3.

SEEKING ADVICE FROM PROFESSIONALS

You can ease the funding search process by utilizing the knowledge and experience of professionals and others who have been successful in obtaining research funding.

University research offices

If you work for (or have very close ties with) a university, contact the research office for advice. Experienced members of staff will be able to direct you to useful sources of funding and offer advice about the grant application process. They can also advise about collaboration projects between departments, between universities and with industry. Therefore, if you work in industry and have an idea for collaboration, or a knowledge transfer partnership, the research office at your chosen university should be able to help. More information about the work of university research offices is provided in Chapter 13.

Experienced colleagues

Experienced and friendly colleagues can be a good source of advice and tips, for example about which organizations to approach and which to avoid. They will also be able to advise you about how to make a successful application (see Chapter 17). If you are an early-career researcher, or a postgraduate student, seek advice from your supervisor or team leader. They might also be able to look through your application and offer feedback before you submit. More information about choosing the right colleagues to work with is offered in Chapter 13.

NETWORKING OPPORTUNITIES

Use networking to help you find out more about funding for your research. This can be done through the use of social media, blogs, forums and micro-blogging sites. Also, most of the larger funding organizations have a twitter feed and email alerts to let you know about new funding opportunities, so sign up to all relevant sites so that you can receive new information as soon as it becomes available.

The Technology Strategy Board

Networking opportunities (and partnerships) are available in the

UK through the following TSB schemes. More information can be obtained from www.innovateuk.org.

- **Catapult Centres** A network of technology and innovation centres that help businesses transform ideas into new products and services. It is possible for you to access equipment and specialist facilities at these centres to test ideas. Find out more about the Catapult programme by visiting www.catapult.org.uk.

- **_connect** An online business networking and open innovation portal. You can meet and network with other organizations to find business partners for collaborative projects, for example. You can also use a range of online tools and services to help with your business ideas and research. Register for free at connect.innovateuk.org/web/guest/home.

- **Innovate UK** An annual networking, conference and exhibition event for business. It brings together a wide range of people from UK and international business, government, academia and the third sector, and can help you to form networks that will benefit your research.

- **Knowledge Transfer Network** A national network that brings together businesses of any size, research organizations, universities, technology organizations and government, to consider such issues as innovation, policy and finance. You can receive advice on funding opportunities through this network. For more information visit connect.innovateuk.org/ knowledge-transfer-networks.

- **Knowledge Transfer Partnerships** These offer businesses the chance to work with academic partners to access knowledge, technology and skills to help improve their business, service or product. The Government funds part of the costs for the partnership, which is paid through the academic partner. More information can be obtained from www.ktponline.org.uk.

- **Themed Missions** This scheme enables innovative businesses to travel overseas to connect with businesses, investors, suppliers and customers. Eligibility criteria depend on specific Missions, with companies selected through a competitive process. More information about the Missions, and the countries to which you can travel, is available at www.innovateuk.org/missions.

If you are an industry researcher in South Africa, similar networking opportunities are available through the Technology and Human Resources for Industry Programme (THRIP) (thrip.nrf.ac.za). Canadian researchers can access networking opportunities through the Industrial Research Assistance Program (www.nrc-cnrc.gc.ca). German researchers can find out more from the Association of German Chambers of Industry and Commerce (www.dihk.de).

PROFESSIONAL BODIES

Networking opportunities can increase significantly if you join the relevant professional body. Other members may have useful advice to offer about finding and applying for funding for research. Also, professional association websites contain helpful information, often with links to funding sites. Professional bodies that may be of interest include:

United Kingdom

- The Association of Research Managers and Administrators www.arma.ac.uk

- The British Educational Research Association www.bera.ac.uk

- The British Psychological Society www.bps.org.uk

- The British Sociological Association www.britsoc.co.uk

- The New Engineering Foundation: The Innovative Institute www.thenef.org.uk

- The Social Research Association the-sra.org.uk

United States of America

- American Educational Research Association www.aera.net
- American Psychological Association www.apa.org
- Association of Clinical Research Professionals www.acrpnet.org
- National Association for Research in Science Teaching www.narst.org

Canada

- Canadian Association for Health Services and Policy Research www.cahspr.ca
- Canadian Association of University Research Administrators www.caura-acaru.ca
- Canadian Educational Researchers' Association www.csse-scee.ca
- Clinical Research Association of Canada www.craconline.ca

Australia

- Association of Australian Medical Research Institutes www.aamri.org
- Association for Qualitative Research aqr.org.au
- Australasian Science Education Research Association (AUS and Asia) asera.org.au
- Australian Association for Research in Education www.aare.edu.au
- Australian Market and Social Research Society www.amsrs.com.au
- Australian Screen Production Education and Research Association www.aspera.org.au
- Australian Vocational Education and Training Research Association avetra.org.au
- Research Australia www.researchaustralia.org

New Zealand

- Health Services Research Association of Australia and New Zealand www.hsraanz.org
- Independent Research Association of New Zealand www.iranz.org.nz
- New Zealand Association of Clinical Research www.nzacres.org.nz
- New Zealand Association for Research in Education www.nzare.org.nz
- New Zealand Heavy Engineering Research Association www.hera.org.nz

India

- Agricultural Economics Research Association www.aeraindia.in
- All India Association for Educational Research www.aiaer.net
- Operational Research Society of India www.orsi.in

Ireland

- Irish Research Staff Association www.irsa.ie

Portugal

- National Association of Science and Technology Researchers anict.pt

The Arab region

- The Arab Science and Technology Foundation www.astf.net

Global

- International Consortium of Research Staff Associations icorsa.org
- World Association of Young Scientists ways.org

FINDING PUBLIC AND PRIVATE FUNDING SOURCES

There are many public and private funding sources for research throughout the world, and contact details for some of the main ones are given below. All these organizations offer small and large research grants, but will have strict rules and regulations attached to their funding, so make sure that you are familiar with these before you submit an application (see Chapter 14). To find out about other grant-awarding bodies, use the databases and directories listed above and in Appendices 2 and 3.

United Kingdom

Government funding

- Arts and Humanities Research Council www.ahrc.ac.uk
- Biotechnology and Biological Sciences Research Council www.bbsrc.ac.uk
- Economic and Social Research Council www.esrc.ac.uk
- Engineering and Physical Sciences Research Council www.epsrc.ac.uk
- Medical Research Council www.mrc.ac.uk
- Natural Environment Research Council www.nerc.ac.uk
- Science and Technology Facilities Council www.stfc.ac.uk
- The British Academy www.britac.ac.uk
- The Higher Education Academy www.heacademy.ac.uk
- The Royal Academy of Engineering www.raeng.org.uk
- The Royal Society royalsociety.org
- The Technology Strategy Board www.innovateuk.org

Private funding

- Cancer Research UK www.cancerresearchuk.org/ funding-for-researchers

- Joseph Rowntree Foundation www.jrf.org.uk/funding

- Nuffield Foundation www.nuffieldfoundation.org

- Sutton Trust www.suttontrust.com

- Wellcome Trust www.wellcome.ac.uk/funding

United States of America

Federal funding

- Agency for Healthcare Research and Quality www.ahrq.gov/funding

- Centers for Disease Control and Prevention www.cdc.gov/about/business/funding.htm

- National Endowment for the Humanities www.neh.gov/grants

- National Institutes of Health grants1.nih.gov/grants

- National Science Foundation www.nsf.gov

- US Department of Education www.ed.gov

Private funding

- Alzheimer's Association www.alz.org

- American Cancer Society www.cancer.org

- American Federation for Aging Research www.afar.org/research/funding

- Compton Foundation www.comptonfoundation.org

- Muscular Dystrophy Association mda.org/research

- National Parkinson Foundation www.parkinson.org

- Robert Wood Johnson Foundation www.rwjf.org_

- Rockefeller Foundation www.rockefellerfoundation.org/grants

- Spencer Foundation www.spencer.org/content.cfm/research

Canada

Federal funding

- Canada Foundation for Innovation www.innovation.ca

- Canada Research Chairs program www.chairs-chaires.gc.ca

- Canadian Institutes of Health Research www.cihr-irsc.gc.ca

- National Research Council Canada www.nrc-cnrc.gc.ca

- Natural Sciences and Engineering Research Council www.nserc-crsng.gc.ca

- Social Sciences and Humanities Research Council www.sshrc-crsh.gc.ca

Private funding

- Alzheimer Society of Canada www.alzheimer.ca

- Burroughs Wellcome Fund www.bwfund.org (US and CAN)

- Cancer Research Society www.src-crs.ca

- Heart and Stroke Foundation of Canada www.hsf.ca/research

Australia

Public funding

- Australian Research Council www.arc.gov.au

- Commonwealth Scientific and Industrial Research Organization www.csiro.au

- National Health and Medical Research Council www.nhmrc.gov.au

- Office for Learning and Teaching www.olt.gov.au

Private funding

- Diabetes Australia Research Trust www.diabetesaustralia.com.au

- Heart Foundation of Australia www.heartfoundation.org.au

- The Australian Cancer Research Foundation acrf.com.au

- The Myer Foundation www.myerfoundation.org.au

Germany

- German Research Foundation www.dfg.de

France

- The French National Research Agency
  www.agence-nationale-recherche.fr

The Commonwealth

- The Association of Commonwealth Universities
  www.acu.ac.uk

- The Ramsay Memorial Fellowships Trust
  www.ucl.ac.uk/ramsay-trust

US–UK

- Fulbright Commission www.fulbright.org.uk

- Marshall Scholarships www.marshallscholarship.org

Worldwide

- Bill and Melinda Gates Foundation www.gatesfoundation.org

- Chevening Scholarships www.chevening.org

- TDR/World Health Organization www.who.int/tdr/en

TIP: GATEWAY TO RESEARCH

Research Councils UK has launched a web portal that provides access to information about 42,000 research projects that have been funded by the seven research councils and the Technology Strategy Board. It is called Gateway to Research and can be accessed at gtr.rcuk.ac.uk. This portal enables you to find out what projects have been supported and how much they have received. You can also find out about research output and impact, all of which will help with your funding search and grant application.

SUMMARY

Although searching for research funding can be a long and laborious process, your search can be made easier by seeking advice from professionals who know about funding opportunities, such as experienced colleagues or staff in university research offices. It is also important to network as much as possible, using formal networks such as those developed by professional associations and the TSB, and informal networks that build up through membership of associations, blogs, social media sites or forums, for example. The websites of public and private funding sources will provide more information about funding opportunities.

Chapter 3

Funding for Academic Researchers

This chapter provides a snapshot of specific grants that are available for academic researchers, organized by subject area. Each entry provides a short summary of the grant, along with practical information about the amount of money available, eligibility criteria and application procedure. Most of these grants are available to researchers of all nationalities, although most funding criteria will specify that grant holders must be employed in a particular country.

The information given is correct at the time of writing (spring 2014), but grants and eligibility criteria can change, so you are advised to seek further information about any grant that is of interest to you. If you cannot find a suitable funding organization in this chapter, details of alternative organizations are listed in Appendices 2 and 3.

FUNDING FOR ALL SUBJECTS

Future Fellowships: Australian Research Council
(AUS and international researchers)

Summary   Set up by the Australian Government to promote research in areas of critical national importance by giving outstanding researchers incentives to conduct their research in Australia.

Amount　　This scheme provides a four-year fellowship. In addition, the ARC may award the administering organization up to $50,000 of non-salary funding per year. This can be used for infrastructure, equipment, travel and relocation costs directly related to the Future Fellow's research.

Eligibility　　Outstanding Australian and international mid-career researchers are eligible to apply (researchers do not have to be permanent residents of Australia). Medical and dental research is not included.

Application　A proposal may only be submitted by an eligible organization through its research office (see the website below for a list of eligible organizations).

Information　To find out more visit www.arc.gov.au. Other funding opportunities include Australian Laureate Fellowships and Discovery Early Career Researcher Awards.

College-Industry Innovation Fund (CAN)

Summary　　This scheme 'provides colleges with funds to acquire significant research infrastructure that will enable partnerships with the private sector to support business innovation'.

Amount　　Up to 40 per cent of the eligible costs of a funded project.

Eligibility　　The fund aims to support partnerships between colleges and the private sector, but will also consider participation from other sectors, such as public and non-profit sectors. It is available for projects in the natural and social sciences, engineering, humanities and health sciences.

Application   Colleges can submit one proposal per competition for each stream. When applying, you will be required to demonstrate clearly how the requested infrastructure will fulfil the objectives of the programme. These objectives can be found on the Innovation website (see below).

Information   To find out more visit www.innovation.ca.

LIFE AND PHYSICAL SCIENCES FUNDING

Wolfson Research Merit Awards: The Royal Society (UK, all nationalities)

Summary   These grants are for outstanding scientists (of any nationality) holding a permanent post at a UK university. They offer a five-year salary enhancement to help recruit and retain researchers in the UK.

Amount   Usually in the range of £10,000 to £30,000 per annum, for up to five years (on top of basic salary).

Eligibility   All areas of the life and physical sciences, including engineering, but excluding clinical medicine. You must hold a permanent post at a university in the UK or have received a firm offer to take effect from the start of the award, and your basic salary must be wholly funded by the university.

Application   You cannot apply until the university vice chancellor and your elected representative have discussed eligibility with The Royal Society Grants Office. Once your application is received it is reviewed by members of the Wolfson Research Merit Awards Panel and then by two independent referees suggested by the panel members. The selection panel considers each nomination and sends their recommendations to the Wolfson Foundation for approval.

Information   To find out more visit royalsociety.org or email grants@royalsociety.org. Other grants include the University Research Fellowship (for early-career researchers), the Newton International Fellowship (for non-UK scientists who wish to conduct research in the UK) and Industry Fellowships (for collaborative projects with industry: see Chapter 4).

FUNDING FOR MEDICAL, BIOLOGICAL OR VETERINARY RESEARCH

David Sainsbury Fellowship Scheme: National Centre for the Replacement, Refinement and Reduction of Animals in Research (UK, all nationalities)

Summary   The NC3Rs 'is an independent, Government-support centre that funds and drives science and technological developments to replace or reduce the need for animals in research, and leads to welfare improvements where animals continue to be used'. The David Sainsbury Fellowship Scheme supports the training and career development of exceptional early-career researchers.

Amount   Five awards are available each year, with a value of £65,000 per annum for three years.

Eligibility   Researchers with up to five years post-doctoral experience at the time of application are eligible to apply. Applications will also be accepted from final-year PhD students, but they must have been awarded their PhD before the Fellowship can commence. You are not eligible to apply if you already hold a permanent contract of employment at the host institute.

Application   Applicants must contact the NC3Rs office to discuss their proposal prior to submission: fellowships@ nc3rs.org.uk. Applications will need to be submitted via an eligible host research organization and a letter of support is required from a head of department, guaranteeing the Fellow space and facilities for the duration of the award.

Information   To find out more visit www.nc3rs.org.uk. Other grants available include project grants, pilot study grants, strategic awards and studentships.

SOCIAL SCIENCES FUNDING

Urgency Grants: the Economic and Social Research Council (UK, international co-investigators)

Summary   A pilot scheme that offers grants for research into 'immediate, rare and unfolding events'. The grants will run for two years enabling researchers to collect data and carry out an initial data analysis. The scheme began in 2013 and will be reviewed on a regular basis.

Amount   Up to £200,000.

Eligibility   You can apply for this grant only if relying on other awarding bodies would result in a missed opportunity to undertake valuable research. The grant will not be available to fund studentships or equipment. Researchers have to be based at a UK research organization that is eligible for research council funding, although international co-investigators can be included in this scheme (see Appendix 1).

Application   You will need to submit a project outline within four weeks of the beginning of the event in question by email to urgentinviteresearch@esrc.ac.uk. If your outline is approved you will need to submit a full proposal within a further four weeks. Your research must begin within one month of grant acceptance.

Information   More information can be obtained from the ESRC website (www.esrc.ac.uk) or by email (urgentinviter-esearch@esrc.ac.uk). Other grants include Research Grants (open call), Professional Fellowships, Placement Fellowships (for knowledge exchange with community groups, for example) and the Future Leaders Research Scheme (for outstanding early-career researchers). If you are a researcher in Hong Kong you may be interested in the ESRC and Research Grants Council of Hong Kong Bilateral Award, for collaboration between social scientists in the UK and Hong Kong.

FUNDING FOR EDUCATION RESEARCH
Education and Social Opportunity Grants:
the Spencer Foundation (US and international)

Summary   Education and social opportunity is one of five areas of enquiry identified by the Spencer Foundation. Small and major grants are offered to researchers conducting studies in this area. Proposals are accepted from the US and internationally, although all proposals must be submitted in English and budgets must be proposed in US dollars.

Amount   Small grants up to $50,000 and major grants of $50,001–500,000.

Eligibility   The principal investigator must be affiliated with a college, university, research facility, school district or cultural institution that is willing to serve as the fiscal agent if the grant is awarded. Research grant proposals from individuals are not eligible. Researchers must have an earned doctorate in an academic discipline or professional field, or appropriate experience in an education research-related profession.

Application   Preliminary proposals must be submitted electronically using the Spencer Foundation's online system. Application guidelines are available on the website (see below).

Information   To find out more visit www.spencer.org. Other areas of enquiry are organizational learning; purposes and values of education; teaching learning and instructional resources; and field-initiated proposals.

ARTS AND HUMANITIES FUNDING
Responsive Mode Research Scheme:
Arts and Humanities Research Council (UK)

Summary   Provides a variety of grants for high-quality research in any subject area within the Arts and Humanities Research Council's (AHRC) remit. Each type of grant has different aims, limits and durations, although many standard rules apply to all. Various grants are available through this scheme, including postgraduate studentships, early career grants and large-scale collaborative research grants.

Amount   Varies, depending on type of grant.

Eligibility   The principal investigator must be resident in the UK. The AHRC piloted a policy to allow international researchers to act as co-investigators on some of its schemes until the end of 2014, when it was reviewed. Proposals may only be submitted by higher education institutions, independent research organizations and Research Council institutes. A list of organizations recognized by the AHRC is available on their website (see below). Museums, galleries, libraries and archives can apply to AHRC in collaboration with a UK university or AHRC-recognised independent research organization (see Appendix 1).

Application   You can apply at any time of the year. All proposals must be completed and submitted via the Research Council's Joint Electronic Submission System (je-s. rcuk.ac.uk). Help and support for applications can be obtained by email (jeshelp@rcuk.ac.uk) or by telephone (01793 444164).

Information   More information can be obtained from the AHRC website (www.ahrc.ac.uk/Funding-Opportunities/ Research-funding) or by email (GrantsPreAward@ ssc.rcuk.ac.uk). Other schemes include the Early Career Research Grant Scheme, the Research Networking Scheme (for the exchange of ideas) and the Follow-on Funding Scheme (to support innovative and creative engagements and enhance impact). Responsive Mode Funding is available from other research councils (see Appendix 4).

HEALTH SERVICE RESEARCH FUNDING
Health Services and Delivery Research Programme: National Institute for Health Research (UK)

Summary   The National Institute for Health Research (NIHR) Health Services and Delivery Research (HS&DR) Programme funds research to produce evidence on the quality, accessibility and organization of health services in the UK. Funding opportunities include commissioned calls, researcher-led calls and themed calls.

Amount   Varies, depending on the type of research.

Eligibility   Research into any aspect of health service quality, accessibility and effectiveness, as long as its importance to the NHS can be demonstrated clearly. The programme will not support basic scientific or epidemiological research on the causes of disease, the testing of new health technologies or diagnostic techniques, or PhD studentships.

Application   Applications are made via an online application form. An outline application is submitted for review. All shortlisted applicants will be invited to submit a full proposal and will normally have eight weeks to do this.

Information   To find out more visit www.nets.nihr.ac.uk. Other funding opportunities include the Health Technology Assessment Programme, the Public Health Research Programme, the Efficacy and Mechanism Evaluation Programme, the Systematic Reviews Programme and the NIHR Clinical Trials Unit Support Funding.

NATURAL ENVIRONMENT RESEARCH FUNDING
Catalyst Grant: Natural Environment Research Council (UK)

Summary   Catalyst Grants are aimed at enabling researchers to develop research partnerships across disciplines and research strategies with the potential for national and/or international impact. These grants allow

researchers to build partnerships and to develop proposals. Catalyst Grants are not intended to support
research projects, but may support some preliminary
research activity. The NERC invites calls for Catalyst
Grants on various subjects.

Amount   Varies, depending on the particular call.

Eligibility   UK researchers normally eligible for funding from
NERC. See Appendix 1 and visit the NERC website
for eligibility criteria (see below).

Application   Applications must be submitted using the Research
Council's Joint Electronic Submission system (je-s.
rcuk.ac.uk).

Information   To find out more visit www.nerc.ac.uk/funding.
Other grants available from the NERC include fellowships and studentships, research programmes
and responsive research awards.

TIP: CROWDFUNDING FOR
POSTGRADUATE STUDENTS

If you are a postgraduate research student it can be difficult to obtain
funding. However, crowdfunding may provide the answer. A new
platform has been set up recently to help students to raise cash for
their tuition and maintenance fees and for research costs. For more
information about this service visit StudentFunder (www.studentfunder.
com). For more information about crowdfunding, see Chapter 6.

SUMMARY

A wide variety of grants is available for academic researchers throughout the world. Most of these are open to researchers of all nationalities, although many will stipulate that researchers must live and work in the country that provides the grant. There are, however, opportunities for collaboration projects for researchers from different countries. All funding has strict eligibility criteria in terms of the type of researcher and research organization that will be supported by these projects.

Chapter 4

Funding for Private Sector and Industry Researchers

This chapter provides a snapshot of specific grants that are available for private sector and industry researchers (some of these are relevant to academic researchers as they provide the opportunity for collaboration between universities and industry). Each entry provides a short summary of the grant, along with practical information about the amount of money available, eligibility criteria and application procedure.

The information given is correct at the time of writing (spring 2014), but grants and eligibility criteria can change, so you are advised to seek further information about any grant that is of interest to you. If you cannot find a suitable funding organization in this chapter, Appendices 2 and 3 list details of alternative sources.

THE SMALL BUSINESS RESEARCH INITIATIVE (UK)

Summary   This scheme does not provide a grant, but instead provides 100 per cent funding through a contract to small businesses in the UK. A government department or public body identifies a specific challenge that is then turned into an open competition for new technologies.

Amount   Varies, depending on the competition.

Eligibility   Small or medium-sized enterprises (SMEs) that
               are working on the development of an innova-
               tive process, material, device, product or service.
               Contracts can only be awarded to legal entities.
               Registered charities are equally eligible to enter
               Small Business Research Initiative (SBRI) competi-
               tions via their trading company limited by guarantee.
               If universities want to apply they must be able to
               demonstrate a viable route to market.

Application   You must register with the SBRI before you
               make an application. Companies with potentially
               interesting technologies and ideas submit an
               application, either through the Technology
               Strategy Board (TSB) or direct to the relevant
               department, depending on the competition. All
               submitted ideas are assessed, and those judged to
               be the most promising are awarded development
               contracts for the first feasibility stage. Some ideas
               may be awarded a second feasibility contract to
               continue with the technology.

Information   To find out more visit www.innovateuk.org,
               email competitions@innovateuk.org or telephone
               0300 321 4357.

HORIZON 2020 (EU AND INTERNATIONAL)

Summary   The EU's new programme for research and innova-
               tion, replacing the Seventh Framework Programme
               for Research and Technological Development (FP7).
               It will run from 2014 until 2020 with a budget of
               nearly €80 billion. The scheme aims to strengthen

EU science, industrial leadership and innovation and address societal challenges such as climate change and renewable energy.

Amount   Varies, depending on current calls and the type of research and innovation.

Eligibility   SMEs will be encouraged to participate across the Horizon 2020 programme. It will be open to organizations from across the world and European researchers will be free to cooperate with their third-country counterparts (researchers and SMEs from any country that is not one of the twenty-eight EU member states or three EEA-EFTA states) on topics of their own choice.

Application   Varies, depending on the funding programme.

Information   To find out more visit ec.europa.eu/research/ horizon2020. The small businesses portal provides information about EU funding and policy and is open to businesses that are based both within and outside the EU (ec.europa.eu/small-business).

INDUSTRIAL RESEARCH ASSISTANCE PROGRAM (CAN)

Summary   Provides financial support to qualified small and medium-sized enterprises in Canada to help them undertake technology innovation.

Amount   Varies, depending on the type of technology innovation.

Eligibility   Businesses must be incorporated and profit-oriented SMEs in Canada with 500 or fewer full-time equivalent employees. The business must have the 'objective to grow and generate profits through development and commercialization of innovative, technology-driven new or improved products, services, or processes in Canada'.

Application   You should contact one of the Industrial Technology Advisors (ITAs) located across Canada, by calling the toll-free number (see below).

Information   To find out more visit the National Research Council Canada website (www.nrc-cnrc.gc.ca), or call the toll-free number to speak to an advisor (1-877-994-4727).

INDUSTRY FELLOWSHIP: THE ROYAL SOCIETY (UK)

Summary   For scientists in industry who want to work on a collaborative project with an academic organization (and for academic scientists who want to work on a collaborative project with industry). The aim is to enhance knowledge transfer in science and technology between those in industry and those in academia in the UK.

Amount   Provides a basic salary for the researcher and a contribution towards research costs.

Eligibility   Covers all areas of the life and physical sciences, including engineering, but excluding clinical medicine. Applicants must have a PhD or be of equivalent standing in their profession and hold a permanent post in industry in the UK (or a university or not-for-profit research organization).

Application   Applications are made online via egap.royalsociety. org. Guidance notes and terms and conditions can be downloaded from The Royal Society website (see below).

Information   To find out more visit royalsociety.org/grants/ schemes/industry-fellowship or email grants@roy-alsociety.org.

BIOMEDICAL CATALYST:
TECHNOLOGY STRATEGY BOARD (UK)

Summary   An 'integrated translational funding programme' in the UK that can support innovative ideas that demonstrate the potential to provide significant positive healthcare and economic impact. The scheme is operated in conjunction with the Medical Research Council (MRC). Three categories of grant are available: feasibility award, early-stage award and late-stage award.

Amount   £180 million of life sciences funding is available. The amount of award varies, depending on the type of grant.

Eligibility   Innovative businesses (small and medium-sized enterprises) and researchers looking to work either individually or in collaboration to develop solutions to healthcare challenges.

Application   If your application is led by a business, you should apply via the TSB website (see below). If your application is led by an academic, you will need to apply on the MRC website (www.mrc. ac.uk). Applications are accepted on a rolling basis for assessment by independent experts. Check the website to ensure that you have accessed the most up-to-date information before you submit your application.

Information　To find out more visit www.innovateuk.org, email competitions@innovateuk.org or telephone 0300 321 4357.

COLLABORATIVE R&D: TECHNOLOGY STRATEGY BOARD (UK AND OVERSEAS)

Summary　Encourages businesses and researchers in the UK to work together on innovative projects in strategically important areas of science, engineering and technology; co-funds partnerships between businesses and between businesses and academia. Frequent competitions for collaborative R&D project funding are held, in a wide range of areas covering specific technical or societal challenges.

Amount　Varies, depending on the type of project.

Eligibility　Projects must be collaborative and business-led but they can partner with academics and research and technology organizations (RTOs) as well as other companies. Specific eligibility criteria for each competition may apply. Although most competitions are for projects based in the UK, some do have scope to work with organizations from other countries.

Application　You must register before you apply. This will enable you to obtain all the relevant supporting information and guidance notes. Competitions usually comprise of an application and assessment process with selected proposals being invited for presentations to a panel of assessors, the TSB and other key stakeholders. Applications are assessed on individual merit by an independent panel of experts.

Information   To find out more visit www.innovateuk.org, email competitions@innovateuk.org or telephone 0300 321 4357.

FEASIBILITY STUDIES: TECHNOLOGY STRATEGY BOARD (UK)

Summary   A single-company or collaborative R&D grant scheme that allows businesses the opportunity to test an innovative idea. Feasibility studies are a way for companies to carry out exploratory studies that could lead to the development and marketing of new products, processes, models, experiences or services.

Amount   Varies, depending on the type of project.

Eligibility   Competitions are open to all UK-based companies and research organizations. Proposals must be collaborative and business-led.

Application   You are required to register on the TSB website before you apply for a competition call. Also, the TSB holds events in support of most competitions and you are advised to attend these events prior to making your application. All application deadlines are displayed clearly on the TSB website (see below).

Information   To find out more visit www.innovateuk.org, email competitions@innovateuk.org or telephone 0300 321 4357.

INNOVATION VOUCHERS: TECHNOLOGY STRATEGY BOARD (UK)

Summary   Designed to help businesses seek outside expertise. Start-ups and small and medium-sized businesses from across the UK can apply for an innovation voucher. All types of innovation are funded.

Amount   Up to £5,000 is available to businesses to work with a supplier for the first time and is used to pay for knowledge or technology transfer from that supplier.

Eligibility   Open to micro, small and medium-sized businesses (SMEs) only, as defined by the EC. Your business must also be eligible to receive grant funding under EC de minimis state aid regulations. There is an upper limit of €200,000 for all de minimis state aid provided to any one business over a three-year period. More information can be obtained from www.gov.uk/state-aid.

Application   A step-by-step guide to completing an application is available on the Innovation Vouchers website. Vouchers are allocated once every three months and closing dates are displayed on the website (see below).

Information   To find out more visit vouchers.innovateuk.org.

SMART: TECHNOLOGY STRATEGY BOARD (UK)

Summary   Offers grants to SMEs for innovative projects with high potential. Three types of grant are available: proof of market, proof of concept and development of prototype.

Amount   Maximum grant is £25,000 and offers up to 60 per cent of the total project cost.

Eligibility   Offers co-funding to UK-based pre-start-ups, start-ups, micro businesses and SMEs, to carry out science, engineering and technology R&D projects that could lead to successful new products, processes and services.

Application   There are six application rounds per year, although you can apply at any time. Once your application is submitted, it is forwarded to independent assessors to begin the assessment process. You should expect to hear from the TSB approximately one month after the close date with respect to its funding decision.

Information   To find out more visit www.innovateuk.org, email support@innovateuk.org or telephone 0300 321 4357.

GOVERNMENT FUNDING FOR RESEARCH AND DEVELOPMENT IN THE UK

According to the National Audit Office, research and development undertaken by public research institutions (research councils and research bodies associated with government departments) fell by £559 million in real terms between 1995 and 2011, whereas government funding of UK business increased by £255 million. For the years 2007 to 2011, government funding of UK business research and development increased at the rate of 37 per cent, and the NAO speculates that this may be to compensate for the reduction in funding that UK business received from non-government sources over that period (due to the economic recession). For more information about these statistics see Research and Development Funding for Science and Technology in the UK, National Audit Office, available for download from www.nao.org.uk.

SUMMARY

National governments are keen to encourage product, process and service innovation, entrepreneurial capacity and environmental sustainability. To do this they provide grants for researchers from private industry and public research organizations to collaborate with projects, exchange expertise, test new ideas and take products to market. Such grants are available throughout the industrial world.

Chapter 5

Funding for Charity, Community and Not-For-Profit Researchers

This chapter provides a snapshot of specific grants that are available for charity, community and not-for-profit researchers (some of which are available for collaboration projects with industry and/or academia). Each entry provides a short summary of the grant, along with practical information about the amount of money available, eligibility criteria and application procedure. Most of these grants are available to researchers of all nationalities, although some funding criteria will specify that grant holders must be employed (and/or conducting research) in a particular country.

The information given is correct at the time of writing (spring 2014), but grants and eligibility criteria can change, so you are advised to seek further information about any grant that is of interest to you. If you cannot find a suitable funding organization in this chapter, refer to Appendices 2 and 3 for details of alternative organizations.

WORLD HEALTH ORGANIZATION REGIONAL OFFICE FOR AFRICA SPECIAL PROGRAMME FOR RESEARCH AND TRAINING IN TROPICAL DISEASES: SMALL GRANTS PROGRAMME (GLOBAL)

Summary   Offers small grants for 'implementation research in disease prevention and control and health systems research'. Proposals are invited from researchers and projects are selected on a competitive basis.

Amount   Up to US$15,000.

Eligibility   Researchers and health professionals should work in disease control programmes of ministries of health, other health sector partners, universities, research institutions and non-governmental organizations. Research projects must have clear objectives and should be completed within twelve months. Priority will be given to projects involving junior researchers and females.

Application   Your proposal must be submitted using the form on the website. Applications can be submitted in English, French or Portuguese. Any applicants that do not complete the proposal form fully will not be considered for funding.

Information   To find out more visit www.afro.who.int.

THE LEVERHULME TRUST: RESEARCH PROJECT GRANTS (UK AND OVERSEAS)

Summary   For innovative and original research projects of high quality and potential, on a theme chosen by the applicant. The grant covers salaries and associated costs and is paid directly to the institution at which the applicant is employed.

Amount	Maximum £500,000; grants may be held for up to five years.
Eligibility	Registered charities in the UK that have research capacity equivalent in standing to that of a UK university. It is also open to registered charities of similar standing in countries where the provision of research funding is seriously limited. Principal investigators are not eligible to apply from institutions or organizations in North America or elsewhere in the EU. The Trust will not fund research that is of direct relevance to clinicians, medical professionals and/or the pharmaceutical industry or policy-driven research where the principal objective is to assemble an evidence base for immediate policy initiatives.
Application	Applications can be made at any of time of the year by using the online application system (grants.leverhulme.ac.uk). An invitation to progress to the second stage will be sent out if your outline application is approved.
Information	To find out more visit www.leverhulme.ac.uk/funding.

THE MYER FOUNDATION: EDUCATION SMALL GRANTS PROGRAM (AUS)

Summary	Provides small grants for education projects in Australia. It supports small community-based organizations.
Amount	Maximum $10,000 and generally for a twelve-month period only.

Eligibility   Applications will be accepted for grants for charitable purposes from Australian incorporated organizations that have been endorsed by the Australian Taxation Office (ATO) as Tax Concession Charities (TCC). If your organization is not endorsed as a TCC, but you are an Australian incorporated organization, you may still be eligible to apply as long as The Myer Foundation can be satisfied that the nature of your project is charitable.

Application   Applications are accepted on an on-going basis, with no specific closing dates. Your application and support material must be submitted through the online process (see website, below).

Information   To find out more visit www.myerfoundation.org. au. Other programmes supported by the Foundation include Arts and Humanities, Poverty and Disadvantage and Sustainability and the Environment.

THE JOSEPH ROWNTREE CHARITABLE TRUST (UK AND EU)

Summary   The JRCT generally funds work (including certain types of research) under five headings: peace; racial justice; power and responsibility; Quaker concerns; and Ireland and Northern Ireland. Grants tend to be made to organizations based in Britain, although the Trust will fund some programmes that are based elsewhere in Europe for work at a European level.

Amount   Varies, depending on the project. Recent grants have ranged from £1,500 to £100,000.

Eligibility   The JRCT will not fund medical or academic research, or research that is more theoretical than practical.

Larger, older national charities that have an established constituency of supporters and substantial levels of reserves will not be funded. It is not necessary to be a registered charity to apply to the Trust. However, the Trust can only support work that is legally charitable as defined in UK law. Your project must have national appeal: local projects will not be funded.

Application   You must submit your application online: applications sent by post, email or fax will not be considered. Make sure that you meet all eligibility criteria before applying. If you are in doubt about this you are encouraged to contact the Trust to discuss your work before you make your application.

Information   To find out more visit www.jrct.org.uk or email (a staff list is available on the website).

THE NUFFIELD FOUNDATION: GRANTS FOR RESEARCH AND INNOVATION (UK, EU, COMMONWEALTH, AFRICA)

Summary   Offered mainly for research (usually carried out in universities or independent research institutes) but also made for practical developments or innovation (often in voluntary sector organizations). There are four programmes in this scheme: Children and Families; Education; Law in Society; and Open Door (for projects that improve social well-being, and meet Trustees' wider interests, but lie outside the three programme areas above).

Amount   £10,000–250,000.

Eligibility   The Foundation does not fund the on-going costs of existing work or services, or research that simply advances knowledge. The Foundation does not make grants for the running costs of voluntary bodies but will consider making a contribution to voluntary sector overheads on funded projects. Projects that are funded are normally in the UK, but EU, Commonwealth and African collaboration projects will be considered.

Application   You must submit an outline application. If this is successful, you will be asked to submit a full application. Detailed guidelines are available on the website (see below).

Information   To find out more visit www.nuffieldfoundation.org.

THE ROBERT WOOD JOHNSON FOUNDATION (US)

Summary   Provides grants for projects in the US and its territories that advance their mission to improve the health and healthcare of all Americans. Most grants are awarded through calls for proposals, although unsolicited proposals are considered through the Pioneer Portfolio scheme.

Amount   Varies, depending on the call. However, most grants are in the region of $100,00–300,000 and run from one to three years.

Eligibility   Public charities, public agencies and universities that are tax exempt under section 501 (c)(3) of the Internal Revenue Code. The Foundation does not support foreign organizations or make grants to individuals.

Application   Applications are made online at www.my.rwjf. org. You will need to register your details to use the system. To be considered for funding from the Pioneer Portfolio, you must first submit a brief proposal.

Information   To find out more visit www.rwjf.org.

ECONOMIC AND SOCIAL RESEARCH COUNCIL: RESEARCH GRANTS (OPEN CALL: UK AND INTERNATIONAL COLLABORATION)

Summary   The ESRC funds social science and economic research across a wide range of disciplines. Open call research grants are available in any research area and topic that falls within the ESRC's remit.

Amount   £200,000–£2 million (100 per cent full Economic Cost (fEC)).

Eligibility   Charities and not-for-profit groups are eligible to apply if they 'possess an existing in-house capacity to carry out research that materially extends and enhances the national research base and are able to demonstrate an independent capability to undertake and lead research programmes'. Other eligibility criteria apply (see Appendix 1).

Application   Applications are made online and can be submitted at any time: there are no fixed closing dates. Visit the ESRC website for application guidelines.

Information   To find out more visit www.esrc.ac.uk, email grant-sesrc@ssc.rcuk.ac.uk or telephone 01793 867125. Similar grants are available from the six other research councils in the UK (see Chapter 2 for details).

SUMMARY

This chapter has provided an overview of the grants that are available for charity, community and not-for-profit researchers. These grants are available from both public and private sources which provide small and large amounts of funding for research projects. Each grant has strict eligibility criteria in terms of the type of group or charity that will be funded, so check these carefully before you submit an application.

As we have seen above, funding bodies tend to offer funds only to researchers working for registered charities, research institutes or universities. If you are a self-employed or retired researcher it can be harder to obtain funds, although there is money available.

Chapter 6

Funding for Self-Employed and Retired Researchers

This type of funding can be difficult to obtain because many of the larger funding bodies will not support individual researchers who are not affiliated to a university or organization that has a good research record. Such organizations include respected trusts, foundations, museums or laboratories (these are termed Independent Research Organizations in the UK: see Appendix 1).

However, despite these difficulties, it is possible for self-employed or retired researchers to access funding. For example, you may be able to obtain funds from trusts and charities, from academies and national bodies, from government departments and publishers. It may also be possible to raise capital through crowdfunding or through various types of sponsorship.

FUNDING FROM TRUSTS AND CHARITIES

Some trusts and charities do offer funding for individual researchers. However, you must pay close attention to the eligibility criteria to ensure that neither you (as an individual researcher), nor your project, are excluded from the funding process. Some funding bodies will only fund individual researchers who have 'close ties' to or have a sponsor from a university, and some will not fund research to write a book, film or play (see over). Examples of grants that are available from trusts and charities include:

- **Research Project Grants from The Leverhulme Trust** These grants are for innovative and original research projects of high quality and potential, on a theme chosen by the applicant. You can apply if you are retired but have maintained close links with your university (or organization of similar standing). Funding is available in the UK and in countries where the provision of research funding is seriously limited. To find out more visit www.leverhulme.ac.uk/funding.

- **Grants from The Joseph Rowntree Charitable Trust (JRCT)** Funding is available for certain types of research under five headings: peace; racial justice; power and responsibility; Quaker concerns; and Ireland and Northern Ireland. The Trust makes grants to a range of organizations and to individuals. Although it is not necessary to be a registered charity to apply to the Trust, it can only support work that is legally charitable as defined in UK law. The JRCT will not fund medical or academic research, or research that is more theoretical than practical. It will also not fund your personal income while you research or write a book, film or play. Your project must have national appeal: local projects will not be funded. To find out more visit www.jrct.org.uk.

- **Biomedical science funding from the Wellcome Trust** It is possible for retired researchers to apply for some funding schemes from the Wellcome Trust as an applicant or co-applicant with the support of a sponsor who must be able to guarantee that space and resources will be made available for the project. Also, your sponsor must hold an established post or a Wellcome Trust Principal/Senior Research Fellowship and have tenure beyond the duration of the grant. To find out more visit www.wellcome.ac.uk/Funding/Biomedical-science.

FUNDING FROM ACADEMIES AND PROFESSIONAL BODIES

Funding from academies and professional bodies may be available to you (in some cases you will need to maintain close ties with an academic or scientific institution). Examples of relevant schemes include:

- **Research Grants from The Royal Society** These grants are available for 'non-tenured researchers and retired scientists' in the history of science field, if you work in association with an eligible institution. The scheme provides support for either research in the history of science (up to £15,000) or the publication of scholarly works in the history of science (up to £5,000). To find out more visit royalsociety.org/grants.

- **Research Grants from the British Medical Association** If you are a member of the BMA you may, as a retired or self-employed researcher, be able to access grants if you are able to obtain a signed letter from the organization in which the research will be undertaken, stating permission to carry out the proposed research. Some of the grants are available for research outside the UK. To find out more visit bma.org.uk.

- **The Wyn Wheeler Research Grant from the Fisheries Society of the British Isles (FSBI)** This grant is available specifically for retired members and is open to bids for up to £6,000 per project. The aim of this grant is to provide members with financial support for their continued activity in fish biology following retirement from full-time employment. To find out more visit www.fsbi.org.uk/grants/research-grants.

Contact the relevant national academy or your professional association to find out whether grants are available for you, as a retired or self-employed researcher (see Chapter 2).

FUNDING FROM GOVERNMENT DEPARTMENTS

Some government departments will fund individual researchers. For example, in the UK the Department for Environment, Food and Rural Affairs (DEFRA) will fund individual researchers (and charities, university spin-offs and groups) if they meet the criteria for a small business. These criteria include having less than 250 employees, an annual turnover of no greater than €40 million and being independent, with less than 25 per cent of the business owned by other enterprises that do not conform to the other criteria. For more information visit www.gov.uk/research-funding-from-defra.

Chapter 2 lists the main sources of public funding in a variety of countries. Visit the relevant websites to find out whether you are eligible to apply.

FUNDING FROM PUBLISHERS

If you intend to write a book after you have completed your research, it may be possible to obtain a small amount of funding from a book publisher in the form of an advance against royalties. However, not all publishers provide an advance and, if they do, amounts can be quite small, starting at around £500. The advance will be greater if you are a published, well-respected author and if your research is on a topic that has wide public appeal. This funding method will not pay for your research, but can help with small expenses. Contact your chosen publisher for more information.

RAISING CAPITAL THROUGH CROWDFUNDING

You may be able to obtain funding for your research using crowd-funding (or crowd-sourced fundraising). This is where a number of individuals come together to pool their money to pay for a particular project or venture. There are a variety of schemes available, many of which are internet-based. As a researcher you can propose a project to a crowdfunding organization (referred to as a 'platform'). This proposal is then put to the crowd who donate to the project.

There are different platforms available, with different ways of operating, so do your research to ensure that you put your proposal to a suitable and reliable platform. You also need to consider issues of intellectual property, patents, online abuse and fund exhaustion if you decide to follow this route.

A directory listing information on platforms currently open to fund-raising from individuals and businesses in the UK can be found at www.crowdingin.com. Websites that enable researchers to present their proposed research to the public for donations include Geek Funder (geekfunder.com) and Rocket Hub (www.rockethub.com). See Appendix 4 for details of books about crowdfunding.

CROWDFUNDING FOR A RESEARCH FELLOW

In June 2014, the University of Portsmouth appointed a research associate to investigate this way of raising capital. However, the initial two-year contract will be extended only if the scholar is able to raise money towards the post's salary costs using crowdfunding. Although the scholar is not expected to raise 100 per cent of earnings, he or she must prove that some funds can be raised using this method. This post is part of a project funded by a grant of £750,000 from the Engineering and Physical Sciences Research Council to research the intricacies of crowdfunding.

FUNDING FROM SPONSORSHIP

If you have close ties with a particular company (perhaps as a former employee) and you intend to carry out research that will be of interest to that company, you might be able to negotiate sponsorship. To do this you will need to present a detailed budget (see Part 2) and a well-presented, persuasive and justifiable proposal (see Part 3). Sponsors will want to know how the research will be of benefit to their company and will want assurances that you have the required expertise and motivation to complete the project to their satisfaction.

When receiving sponsorship for research projects, you must be aware of the financial relationship that you have with your sponsor and ensure that this does not in any way affect your research methods and outcomes. Sponsors may feel that, since they have paid for you to conduct the research, they have a say in how it is carried out and how outcomes are reported. This must be resisted, and advice on how to do this is offered in Chapter 21.

If you are interested in obtaining sponsorship for your research, Appendix 4 gives details of books that will provide you with more information. If you are looking for sponsorship for a research degree, visit Scholarship Search (www.scholarship-search.org.uk) and the Scholarship Portal (www.scholarshipportal.eu) to search for funds.

SUMMARY

Individual researchers can find it difficult to obtain funds because many large funding organizations only make grants to researchers who work in universities, registered charities or Independent Research Organizations. However, it is possible to obtain funding if you can prove that you have the required expertise and motivation to complete the research project successfully. Some funding bodies will also request that you have an academic sponsor or maintain close ties with a university or research institute.

Once you are aware of all possible funding organizations, you can go on to choose a funder. To do this you need to ensure compatibility of subject, purpose and ethos, and make sure that you meet all eligibility criteria.

Chapter 7

Choosing a Funding Organization

Now that you are aware of the various types and sources of funding that are available, you can go on to choose the most appropriate funder.

Preparing research proposals, working out budgets and completing application forms is a time-consuming process that, for some researchers, is unpaid. Therefore, you need to ensure that you apply to a funding organization that provides the best chance of success. To do this you must ensure compatibility of subject, purpose and ethos, and make sure that you meet eligibility criteria (this includes researcher and organization eligibility).

ENSURING COMPATIBILITY OF SUBJECT

If you are responding to a call for research, the subject area and topic are specifically defined (see Example 1, below). You must ensure that your proposal fits the call. If in any doubt, contact the funding organization to discuss your ideas and to find out whether it is worth submitting an application. Most will be happy for you to do this: indeed, successful grant applications often result from proactive and effective communication between funder and researcher (see Chapter 17).

EXAMPLE 1: CALL FOR RESEARCH FROM THE ESRC

Europe–China call for collaborative research on The Green Economy and Understanding Population Change

'Applications are invited for joint projects under the following priority themes, addressing key issues where true added value can be gained from collaboration.

The Green Economy

- The 'greenness and dynamics of economies'

- Metrics and indicators for a green economy

- Policies, planning and institutions (including business) for a green economy

- The green economy in cities and metropolitan areas

- Consumer behaviour and lifestyles in a green economy

Understanding Population Change

- Changing life course

- Urbanization and migration

- Labour markets and social security dynamics

- Methodology, modelling and forecasting

- Care provision

- Comparative policy learning

Proposals should include leading European researchers wishing to develop contacts with leading researchers in China, and involve participation from at least two different participating European countries and a Chinese consortium.'

ESRC, www.esrc.ac.uk, October 2013

Other funding organizations set their funding priorities each year. Subjects and topics can change, so you will need to keep abreast of these and make sure that you are up-to-date (see Example 2).

EXAMPLE 2: RESEARCH PRIORITIES FROM THE AXA RESEARCH FUND

Thematic priorities – the AXA Research Fund 'Open Grounds'

'AXA Research Fund's thematic priorities ('Open Grounds') for 2013 will be the following:

Life Risks Because the remarkable gain of about 30 years in life expectancy in developed countries stands out as one of the most important accomplishments of the 20th century, this Open Ground will revolve around the concept of Ageing in the 21st century: from malleability of ageing, to health, vitality and the study of longevity as a new form of social inequality.

Socio-economic Risks Because it is now increasingly possible to track the location, health state, social and economic habits of nearly every person on the planet, and because the flows of Big Data can unleash powerful analytic capabilities, this Open Ground will revolve around the concept of Digital as an On-going Revolution: from privacy issues, to e-health and e-marketing evolutions.

Environmental Risks Because of increases in the frequency and intensity of climate change-induced extreme weather events combined with escalating population density, this Open Ground will revolve around our societies' Ability to Face Extreme Hazards: from the modelling of climate change at small and mid-scale, to socio-economic resilience and the prediction and impact assessment of local, extreme and recurring events.'

AXA, www.axa-research.org, October 2013

Some funding organizations, however, will accept proposals on any subject (see Example 3). When choosing this type of funding organization, you must ensure that you meet all the eligibility criteria, which can be very detailed (see below).

EXAMPLE 3: FUNDING FROM THE LEVERHULME TRUST
Research Project Grants

'The aim of these awards is to provide financial support for innovative and original research projects of high quality and potential, the choice of theme and the design of the research lying entirely with the applicant (the principal investigator). The grants provide support for the salaries of research staff engaged on the project, plus associated costs directly related to the research proposed, and the award is paid directly to the institution at which the applicant is employed.

'Proposals must reflect the personal vision of the applicant and demonstrate compelling competence in the research design. The Trust favours applications that surmount traditional disciplinary academic boundaries and involve a willingness to take appropriate degrees of risk in setting research objectives.'

The Leverhulme Trust, www.leverhulme.ac.uk/funding, October 2013

ENSURING COMPATIBILITY OF PURPOSE

Funding organizations often have strict criteria about the purpose of the research, and, depending on this purpose, could refuse to fund your research. The following list provides examples:

The Leverhulme Trust (www.leverhulme.ac.uk) will not fund:

- applications for research of which advocacy forms an explicit component;
- research that is aimed principally at an immediate commercial application;

- applications in which the main focus is on capacity building, networking, or the development of the skills of those involved.

The Nuffield Foundation (www.nuffieldfoundation.org) will not fund:

- the running costs of voluntary bodies or the continuing provision of a service, however worthwhile;

- the production of films or videos, or for exhibitions;

- research that simply advances knowledge.

The Bill and Melinda Gates Foundation (www.gatesfoundation.org) will not fund:

- projects addressing health problems in developed countries;

- political campaigns and legislative lobbying efforts;

- building or capital campaigns;

- projects that exclusively serve religious purposes.

TIP: KEEP UP TO DATE

Make sure that you obtain the most up-to-date funding guidance notes from each organization as these can change on a regular basis. Ensure that the purpose of your research meets stated criteria.

ENSURING COMPATIBILITY OF ETHOS

When choosing a funding organization, it is important to ensure compatibility of ethos. There are organizations with commercial, political or religious agendas, for example, that can influence the projects that they fund. They can also put pressure on researchers to achieve certain favourable outcomes.

All researchers should adhere to ethical guidelines/policy laid down by professional bodies, governments and well-respected funding bodies. As an ethical researcher you must ensure that your chosen funding organization takes a similar line. Part 4 of this book provides information about acting ethically, addressing conflict of interest and avoiding biased financial relationships. It also offers advice about ensuring that you work with an ethical funding organization.

MEETING ELIGIBILITY CRITERIA

Research funding almost always comes with strict eligibility criteria. It is important to become familiar with these when you begin your funding search to ensure you don't waste time making an application that will not be accepted. In particular, you will need to know about eligibility criteria concerning the following:

- **Required country of domicile** Do researchers need to live in a particular country? Does this apply to all researchers? Some funding organizations may insist that the principal investigator is from a particular country, but co-investigators or collaborators can be from a different (sometimes specified) country.

- **Nationality of researchers** Does the funding organization insist on any particular nationality?

- **Type of organization** Are all organizations funded, or are there strict limits about the type of organization that can apply? Research Council funding in the UK, for example, is only available for higher education institutions, research institutes and named Independent Research Organizations (see Appendix 1).

- **Type of researcher** Can all researchers apply for funding, or is it only available for certain types of researcher? Some funding councils may offer funds to independent researchers only if they have close ties to a larger research organization, such as a university (see Chapter 6). Many organizations will not fund individuals or offer money for personal financial assistance.

- **Place of research** Does the research have to be conducted in a particular country?

- **Research subject area** What subjects are funded? Are any topics within that subject area excluded?

- **Research methods and methodology** Will your methodology or any of your methods be excluded from funding? Some funding organizations will not provide funding for large-scale survey work carried out by outside consultants, and others are reluctant to provide funds for experimental methods that have not been proven.

- **Innovation** Some funding bodies specify that the research must be innovative. Are you able to stress and prove innovation (see Chapters 15 and 17)?

- **Previous or present funding** Most funding bodies will not provide funds for a project that is already funded or for research already done. For example, the National Institutes of Health in the US states that: 'Every project NIH funds must be unique. By law, NIH cannot support a project already funded or pay for research already done, so you cannot have overlap with your other applications.'

TIP: DON'T DISMISS SMALL GRANTS

Don't overlook funding bodies that only offer small grants. It can be easier to obtain a small grant than a large one. It is possible to use small grants for some preliminary research that may help you to secure a larger grant in the future. This will show funding bodies that you have been successful in obtaining a grant and that you have some useful data to inform your future research. This can help to improve your chances of success when applying for larger grants.

SUMMARY

It is important to choose the right funding organization from the outset to avoid wasting time and effort submitting an application that won't be accepted. If you are a contract researcher time spent on unsuccessful proposals is also unpaid. Major reasons for non-acceptance include incompatibility in the subject, purpose and ethos. It is also important to study eligibility criteria carefully to ensure that all these are met before you begin your application.

Once you are sure that you have found an appropriate funder, you can begin to work on your application. The first part of this is costing your project, and advice on how to do this is provided in Part 2.

COSTING PROJECTS

Chapter 8

Knowing about Costing Methods

Once you have identified a potential funder it is important to understand more about the costing method used by the funding organization. Costing methods will vary, depending on the sector, country and funding organization. For example, in the UK the Transparent Approach to Costing and Full Economic Costing are both used in higher education, whereas in the US the National Institutes of Health have developed the Modular Grant Application system for medical research. You will need to understand how to categorize costs and learn which are allowable and unallowable within the relevant costing system.

TRANSPARENT APPROACH TO COSTING (UK)

The Transparent Approach to Costing (TRAC) is the standard method used for costing in higher education in the UK. It has been in use since 2000 for costing the main activities of higher education institutions: teaching, research and other core activities. Universities are required to submit an annual TRAC return to the relevant funding council. Although non-university organizations in the UK are not expected to use this method, funding organizations will want proof that costing methods are robust. Full Economic Costing (see over) is an extension of the TRAC method.

FULL ECONOMIC COSTING FOR RESEARCH (UK)

Full Economic Costing for Research (fEC) was introduced in September 2005 by the UK Government (the abbreviation fEC is used rather than FEC because the latter is recognized as referring to Further Education Colleges). This is a method used by higher education institutions to calculate the cost of research projects, and takes account of all direct costs and associated indirect and estates costs.

This way of calculating the costs of research projects is now standard across the higher education sector in the UK. Your research office will provide further guidance about fEC and help you to calculate the full economic cost of a proposed research project (industry researchers can also use university research offices if working in collaboration with academic researchers). Many universities now provide tools to assist you in building the cost of research projects on a full economic cost basis. Once you have indicated the fEC of a project on your grant proposal, funding bodies then pay a percentage of this sum (80 per cent from research councils for most fund headings). More information about working out these costs is provided in Chapter 9.

However, charities in the UK tend not to fund on a proportion of fEC because they expect general running costs to be provided by Government, through funding to universities. The Government has established a specific revenue stream (the Charity Support Fund) to contribute towards the fEC of research funded by charities at universities in England. This is distributed via the block grant to universities from the Higher Education Funding Council for England (HEFCE). Similar schemes are available in the rest of the UK.

NIH MODULAR GRANT APPLICATIONS (US)

The Modular Grant Application method was introduced in 1998 by the National Institutes of Health (NIH) in the US. The method was designed to 'focus the minds of researchers on science, rather than the intricacies of budget development', and to reduce the time spent between receipt of application and grant award.

Using this method specific modules or increments of $25,000 are established in which direct costs must be requested, with a maximum level for requested budgets of $250,000 direct costs. The Modular Grant Application is used for Research Project Grants, Small Grants, Academic Research Enhancement Awards Grants, Exploratory/Developmental Research Grants and Clinical Trial Planning Grant Programs.

Modular budget guidelines and sample modular budgets are available from the NIH website: grants.nih.gov/grants/funding/modular. If you have any questions regarding this system email the Grants Information Office: GrantsInfo@nih.gov.

CATEGORIZING COSTS

You will need to categorize costs using the relevant costing method. In general, this categorization tends to be into direct costs (staff salaries, equipment and materials) and indirect costs (library facilities and estates). However, using the fEC method in the UK, costs are now categorized into the following three categories:

1 **Directly incurred costs** Items or services incurred or purchased specifically for a project. They include consumables, travel and subsistence, research assistants, dedicated technicians, support staff and equipment purchase.

2 **Directly allocated costs** Costs of services on a project, where the services are shared by other activities and projects. They include major research facilities, estates, investigator's time and laboratory technicians. Different methods can be used to produce estimated costs, although the most common is to produce charge-out rates for each of these items that are generally applied to researcher full-time equivalents (FTEs).

3 **Indirect costs** Costs not directly related to any one research project or activity, but which are a necessary part of the costs of

undertaking the project or activity. They can include staff and non-staff costs in central service departments, academic support time or estates costs that relate to central services. A standard estimate is used to generate these costs.

TIP: WATCH YOUR COSTS

When working out costs, take care to ensure that they appear in only one of the three categories described above. You must ensure that there is no double counting. Direct costs are favourable as there is more scope for transferring money from one budget heading to another (indirect costs cannot be transferred).

It is important that you understand the difference between the cost categories in order to calculate your figures accurately and produce your costs in the format required by funding bodies. Also, as we have seen above, many charities in the UK only fund certain cost categories (see Example 1).

EXAMPLE 1: CANCER RESEARCH UK, DIRECTLY ALLOCATED OR INDIRECT COSTS

'Cancer Research UK will pay the directly incurred costs of research. Cancer Research UK will not pay either directly allocated (including estates costs) or indirect costs on individual research awards. Awards are provided on the understanding that the host institution will meet directly allocated and indirect costs (previously referred to as overhead costs) including lighting, heating, central support staff salaries, costs of equipment maintenance (unless the equipment has been purchased by Cancer Research UK), telephones, photocopying, postage etc. (except in special cases where the volume of paperwork and mailings are considerable, e.g. epidemiological or behavioural studies), use of library facilities and general laboratory and office equipment. Cancer Research UK will consider requests for a contribution to the maintenance costs of the equipment, purchased through a Cancer Research UK award. Where institutions operate a policy of access charges to equipment, Cancer Research UK will consider payment of an access charge in lieu of consideration of maintenance costs. However, having paid for the equipment, in whole or in part, Cancer Research UK will not pay for access under full economic costing. If you are in any doubt as to what might constitute a directly allocated or indirect cost, please contact the office before submitting your application.'

Cancer Research UK, www.cancerresearchuk.org, October 2013

Indirect costs in the US

In the US indirect costs are now known as Facilities and Administration (F&A) costs. All universities that are in receipt of federal funding are required to negotiate a F&A cost rate agreement with their federal audit agency. These rates are generally accepted by commercial and industrial sponsors and must be used when estimating F&A costs in grant proposals. More information about these rates can be obtained from your research administration office. Members of staff will help you to work out the appropriate amounts and will negotiate on your behalf with a prospective sponsor who does not find the federal rate acceptable.

KNOWING ABOUT ALLOWABLE AND UNALLOWABLE COSTS

Within each costing method there are allowable and unallowable costs. Some funding bodies are very specific about what will and will not be funded (having to adhere to a specific costing method), whereas others are more flexible, as illustrated in the examples below.

Federal funding (US)

In the US the Office of Management and Budget (OBM) Circular A-21 (Cost Principles for Educational Institutions) defines allowable and unallowable costs for federally-funded sponsored research awards. This circular identifies four tests of 'allowability':

- They must be reasonable.

- They must be allocable to sponsored agreements under the principles and methods provided in the circular.

- They must be given consistent treatment through application of those generally accepted accounting principles appropriate to the circumstances.

- They must conform to any limitations or exclusions set out in the principles or in the sponsored agreement as to types or amounts of cost items.

Section J of OMB Circular A-21 goes on to discuss unallowable costs. These include:

- advertising and public relations;

- alcoholic beverages;

- alumni activities;

- bad debts;

- certain legal costs;

- charitable contributions;

- contingencies;

- entertainment;

- fines and penalties;

- first-class air travel;

- fundraising;

- investment management;

- goods and services for personal use;

- housing of officers;

- lobbying costs;

- losses on sponsored research agreements;

- memberships in civil, community and social organizations;

- selling and marketing costs;

- telephone line costs.

For more information about the OMB Circular A-21, visit www. whitehouse.gov/omb.

Charity funding

Most charities that offer research grants tend to specify allowable and unallowable expenses and these can vary considerably, depending on the charity. For example, the Wellcome Trust (www. wellcome.ac.uk) will fund salaries, materials and consumables, travel and publication costs, but will not fund estates costs and operational costs. Breast Cancer Campaign (www.breastcancer-campaign.org) will not fund salary recovery costs on tenured posts, publication and printing costs, but will fund laboratory materials and consumables directly attributable to the project.

Once you have found a compatible funding organization, you must ensure that you obtain the most up-to-date guidelines about allowed and disallowed costs. If in any doubt, contact the funding body for clarification (Chapter 17 stresses the importance of establishing a good working relationship with potential funders).

SUMMARY

Knowing about costing methods is crucial to accurate and successful project costing. You cannot work out your costs without first understanding the methods that are adopted by your preferred funding organization. You also need to know how to categorize your costs into direct and indirect costs. Funding organizations have strict rules about what will and what will not be funded and your chances of success will increase significantly if you adhere to these rules and regulations.

Once you have a greater understanding of costing methods, approaches and procedures, you can go on to work out your costs.

Chapter 9

Working out Costs

New or early-career researchers who have not applied previously for a research grant may find working out costs to be a long and laborious process. The information given in this chapter will help to ease the journey by offering advice about sourcing figures for your budget, calculating costs, hiring and paying staff and working within the caps and limitations that are imposed by funding organizations.

SOURCING FIGURES FOR YOUR BUDGET

When producing your research budget it is vital to get your figures right. This shows that you are competent and organized and that you understand the true costs of your research. Funding bodies will scrutinize your figures for accuracy and your application will fail or be returned if your figures are not correct. There are various ways to source figures for your budget:

- university or employer salary scales and benefits database;
- your university research office;
- your estates management department;
- your human resources department;
- your communications/marketing office;
- quotations from internal departments, such as your university printing and binding service;

- quotations from vendors (funding bodies often require up to three quotations, to ensure that you are getting value for money: see Chapter 11);

- tenders or contracts from outside individuals/organizations (these must be open to competition);

- your supervisor or mentor;

- departmental administration office;

- funding bodies (many provide information about salary levels and fringe benefit rates, for example: see below).

CALCULATING COSTS

We know that costs tend to be divided into two categories: direct costs (staff salaries, equipment and materials) and indirect costs (library facilities and estates). When calculating these costs – if you work for a university – consult first with your university research office (or research administration). This is because pre-prepared tools, such as Excel spreadsheets, are available to help you save time and work out exact rates.

Other costs, such as Facilities and Administration (F&A) costs and fringe benefit rates in the US are calculated using a negotiated rate (see Chapter 8 and below). You will need to know these rates before you can calculate costs. Research offices in universities provide a detailed list of all required rates, along with worksheets and scenarios to help you. In some cases, research offices will work out costs for you. However, if you are the principal investigator it is important that you check all costs and calculations: after all, you are the person most familiar with your research.

If you are not a university researcher, seek advice from experienced colleagues and from the funding body. Again, tools are available to help you save much time and effort. Chapter 17 highlights the importance of building a good working relationship with your preferred funding organization: many provide help and advice about costing projects, so make sure that you use all the information and tools available.

HIRING AND PAYING STAFF

Members of staff that could work on a research project include:

- the principal investigator;
- co-investigators;
- collaborators (see Chapter 10);
- joint lead applicants;
- research assistants;
- research students;
- technical, clerical and administrative staff;
- consultants (in the US federal regulations preclude the use of consultants who are employed by the federal government in any capacity).

When working out costs you need to take into account the cost of hiring any new members of staff that are required for the project, salary levels for their grade/expertise and the cost of fringe/employee benefits (see below). Salary costs should include any increments, promotion or re-grading, where appropriate. Your human resource department, finance office or research office will be able to provide information about salary levels, grades, increments, fringe benefit rates and the standard rate for hiring staff. You must also ensure that you work within the salary caps and limitations described below.

TIP: DON'T UNDERESTIMATE STAFF TIME

Researchers who are new to the costing process often underestimate the amount of staff time required to undertake a task (researcher and associated staff time). In these cases it is better to overestimate, rather than underestimate. Seek advice from experienced colleagues if in doubt.

Categorizing staff costs

It can be difficult for new and early-career researchers to understand how to categorize staff costs. The following list provides a general guide for researchers in the UK. However, categorization can be affected by specific projects, country and funding body regulations, so you should seek further advice specific to your needs.

- **Principal Investigator and co-investigator time costs** (calculated from the time that they expect to spend on the project and an appropriate measure of their salary costs for a full working year) are classified as directly allocated costs.

- **Research assistant payroll costs** (calculated from the combined salary, National Insurance and superannuation costs based on the proposed initial spine point, increment date and FTE for the research assistant) are classified as directly incurred costs.

- **Recruitment costs for research staff** are classified as directly incurred costs.

- **Research studentships** are classified as directly incurred costs but are treated as an exceptional cost by the research councils in the UK.

- **The costs of technical, administrative and clerical staff already employed by an organization** (calculated from the time that the member of staff will be required for the project and an appropriate measure of their salary costs) are classified as directly allocated costs.

- **The costs of technical, administrative and clerical staff** hired and employed specifically for the project (calculated from the combined salary, National Insurance and superannuation costs given the proposed initial spine point, increment date and FTE) are classified as directly incurred costs.

Chapter 8 provides more information about categorizing costs into directly incurred costs, directly allocated costs and indirect costs.

TIP: DEPLOYMENT OF STAFF

If multiple staff are to work on your project, funding organizations will want a clear breakdown of how staff will be deployed across the different components of the research programme for the duration of the project.

Fringe benefit rates

'Fringe benefits' is the term used in the US for employee-associated costs such as health plan expenses, pension plan expenses and tax-related expenses, among others. In the UK and other countries these can be termed employee benefits, benefits in kind or perks, for example. Reimbursement for fringe benefit costs, expressed as a percentage of total salaries, is provided by funding bodies. In the US, universities negotiate fringe benefit rates with the federal government on an annual basis. This is applied to salaries funded by government grants and contracts, and to sub-awards from non-governmental sponsors. These rates can provide the largest component of cost on large-scale research projects.

To work out fringe benefit rates, costs are expressed as a rate by employee class (regular, temporary or postgraduate fellow, for example). The rate is the combined costs of these benefits divided by the total salaries in each employee class. These rates are then applied to the relevant employee salary to represent the associated benefits for that type of employee. Your university research office will provide a list of rates and offer more guidance about fringe benefit rates for your grant submission.

The graduate student stipend rate

Stipends are paid to students to cover their living costs. In some countries the national minimum stipends are agreed each year by research councils. Universities can choose to increase the student stipend above the minimum payment if located in an area where the cost of living is high, or if the university is having difficulty recruiting students, for example. Students cannot receive less than the research council specified minimum stipend.

In the UK and elsewhere, stipend rates can also be set by funding bodies so you will need to obtain these rates when you prepare your budget. If you intend to use graduate students on your research project, contact your university research office or chosen funding body for more information about the stipend. As we have seen above, studentships (fees and maintenance) are treated as an exceptional cost by research councils in the UK (research councils will fund 100 per cent of the stipend costs).

WORKING WITHIN CAPS AND LIMITATIONS

The previous chapter stressed the importance of obtaining the most up-to-date funder guidelines concerning approaches to funding, cost categories and information about allowed and disallowed costs. All costs must be produced within these guidelines. Funders can also produce caps on the amount of funding for specific costs or for the whole project and, again, you must work within these caps. A major reason for application failure is that researchers ignore (or misunderstand) these caps.

Caps and limitations under which you may have to work include the following:

- **Salary caps and stipends** In the US salary caps are established by the Department of Health and Human Services (DHHS), which includes agencies such as the National Institutes of Health (NIH), the Department of Defense (DoD) and Food and Drug Administration (FDA). These salary caps are indexed to a specific Government Executive Pay Level and are reviewed periodically. You cannot apply for more than the cap, although it is possible for your institution to pay beyond the cap with non-grant funds. In the UK and elsewhere, there is more flexibility on researcher salaries, although graduate stipend rates are set by research councils (see above). If there is not a cap on salaries you will need to demonstrate that the salary levels sought are commensurate with the skills, responsibilities and expertise necessary to carry out the role required.

- **Equipment purchase** Some funding bodies will place a cap on equipment costs, or will ask that the host institution make a contribution towards equipment that costs over a certain amount. For example, the Wellcome Trust (www.wellcome. ac.uk) expects a contribution of at least 10 per cent on an equipment item costing £100,000 or more, and at least 20 per cent on one costing £500,000 or more. If you hope to receive funding for expensive equipment you will need to provide extra justification for the inclusion of these items (see Chapter 11). Most funding bodies will also request that you obtain at least three quotations from different suppliers.

- **Large surveys by specialist survey companies** Some funding bodies will place a limit on the amount of funding available. Other funding bodies will insist that, if you intend to go down this route, you must ensure that the survey company is subject to external competition so that the best value for money can be achieved (see Chapter 11).

SUMMARY

When applying for funding you need to work out your costs carefully, while paying close attention to funding body regulations, limits and caps. There are various sources of information to help you work out costs, including human resource departments, finance departments and research offices. In some cases members of staff will work out costs for you, but if you are the principal investigator it is imperative that you check all costs carefully as you are the person most familiar with the project.

If your project involves the purchase of expensive equipment and/ or services, you will need to check that there are no limits on these costs. Also, if you are working in collaboration with other research organizations, there are important issues about ownership to address: these are examined in Chapter 10.

Chapter 10

Working with Collaborators

Many researchers work with collaborators. This can include collaboration among individual researchers in centres of excellence or interdisciplinary research groups; between sectors such as academia and industry; and international collaboration between individual researchers, research teams or scientific departments. Collaboration can be encouraged by social, economic, professional, scientific or political factors.

If you decide to work with a collaborator, there are important issues to consider, such as choosing the right collaborator, building a good working relationship, preparing joint budgets and addressing ownership issues.

CHOOSING THE RIGHT COLLABORATOR

When choosing a collaborator, detailed information gathering is vital. Find out all you can about the potential collaborator. What papers have they written? What are their preferred research methods? What have others said about their work? What organization do they work for? What reputation does this organization have?

In addition to these general questions, work your way through the following list. Several 'no' answers should encourage you to think more about whether collaboration is the best way forward and/or whether you have chosen the best collaborator for your project.

The collaboration project   Yes   No

Will the collaboration

- enable the cross-fertilization and generation of ideas?

- enable the sharing of knowledge, techniques and skills?

- enable the transfer of knowledge and skills?

- help with the advancement of knowledge?

- raise the visibility of your work?

- help to raise your research profile?

- help to increase your research output and productivity?

- be mutually beneficial?

- be challenging for all parties?

- be enjoyable for all parties?

- Will funding be available for this project (and extra costs associated with collaboration)?

- Do the advantages of collaboration outweigh the potential disadvantages?

The collaborators

- Do all parties understand the benefits to be gained?

- Do all collaborators have the same aims and objectives for the project?

- Can your collaborators bring useful knowledge, skills or expertise to the project?

- Do the skills and expertise of collaborators complement your own?

- Do all parties have the skills required to work as part of a team?

- Do your collaborators provide a unique perspective (international, for example)?

- Do your collaborators provide the opportunity to extend into other disciplines?

- Can your collaborators supply tools or equipment for the project?

- Will your collaborators adhere to accepted author protocols in published material?

- Can you reach agreement on ownership of intellectual property?

- Can you reach agreement on the sharing of potentially sensitive information?

- Are all parties upfront and realistic about their time expectations?

- Are all parties trustworthy and upfront about their organizations' strategy?

TIP: A SENSE OF PARTNERSHIP

The most successful collaboration projects are those that have developed a real sense of partnership between all members. This partnership often continues once a project has finished, leading to future collaboration projects.

BUILDING A GOOD WORKING RELATIONSHIP

Working through the above list will help you to choose the right collaborator. You then need to build and maintain a good working relationship so that you can prepare a joint budget with which all parties are happy, and so that your project can progress smoothly. The following issues are important when building and maintaining a good working relationship.

- **Management and leadership** If you are the principal investigator you will need to take on the role of manager. You will need an in-depth knowledge of the project and have the ability to drive it forward, managing and leading as appropriate. If you are not the principal investigator you will need to think about who is to take on the management role and whether this person has the required experience and skills.

- **Communication** Face-to-face meetings will need to be conducted on a regular basis. If collaborating with international researchers this needs to be factored into your budget (see over). Video-conferencing and Skype can provide cheaper face-to-face communication if international travel proves to be too expensive. An overall communication routine will need to be established to supplement face-to-face meetings and you will need to ensure that all collaborators have access to required technology. It is also important to encourage personal, informal exchange among team members and collaborators. As a project progresses, collaborators tend to develop a better understanding of the research problem and a common vocabulary to communicate results.

- **Networking** This activity should be encouraged among all collaborators, especially if the collaboration is international or across different sectors. Taking part in a collaboration project can greatly extend a researcher's network and provides the opportunity to raise your research profile. It also enables you to gain skills in presenting complex ideas to a more diverse range of people. More information about building networks is provided in Chapter 2.

- **Understanding** Collaboration projects work better if all members of the team have a clear understanding of their roles and functions from the outset. It is also important to foster social, cultural and organizational understanding if the project is to cross international, social and organizational boundaries. Technical understanding is crucial for all those involved in technical aspects of the project.

TIP: COLLABORATOR COMMITMENT

Ensure that all collaborators understand their commitments from the outset. Collaboration projects can vary considerably: some researchers will have substantial input, whereas others may only be called upon to offer advice. Everyone should understand what is expected from the start so that there are no misunderstandings as the project progresses.

PREPARING JOINT BUDGETS

Extra costs are associated with collaboration projects, and these must be considered when working out your costs for your grant application. These can include:

- **Travel and subsistence costs** For researchers who are located at a distance from each other. Some funding bodies will not pay for travel, so you and your collaborators may have to think of innovative ways to find funds.

- **Costs associated with time** More time may be needed to prepare joint proposals, secure joint funding or write joint papers. More time may also be needed for travel between locations and for communication equipment and activities.

- **Management costs** Larger projects and a greater number of collaborators will require more formal management and could involve the merging of two very different management cultures.

- **Administrative and technical costs** Again, there may need to be
  a merging of very different administrative and technical
  systems, which can take time and, therefore, incur additional
  costs.

When working out costs for joint funding bids it is important to
find out about allowed and disallowed costs (see Chapter 8) and
understand how to categorize the staff costs of all parties in the
collaboration project (see Chapter 9). You must also ensure that
you receive all necessary signatures from collaborators before submitting your funding application (see Chapter 16).

ADDRESSING OWNERSHIP ISSUES

Two main ownership issues need to be addressed when working
with collaborators: ownership of equipment bought with research
grant funding, and ownership of research output.

Equipment

In general, equipment purchased from grant funds for use on the
research project (for which the research grant was awarded) belongs
to the research organization (e.g. university, laboratory or museum).
In some cases, however, the funding body will want to retain ownership of the equipment throughout the period of the grant and
possibly beyond. This will be made clear when funds are granted.
Also, if the equipment is not used as intended, or if use diminishes
over the course of the project, the funding body can request that it
is sold and the proceeds returned to the funding body.

When collaborating with other researchers, ownership issues can
arise over equipment that was bought using funds provided for the
project (if the funding body does not retain ownership). Discuss
these issues with all collaborators before you submit your grant
application so that you can reach agreement about who owns the
equipment during the project and on completion. You will also need

to adhere to caps and limitations placed by funding bodies on the purchase of expensive equipment (see Chapter 9).

Research output

As we have seen in the checklist above, it is important to ensure that your collaborators have the same aims and objectives for the research, are willing to discuss the issue of intellectual property, are willing to share sensitive information and are trustworthy and upfront about the future strategy of their organization.

All these issues affect your research output, including when and how you disseminate results. For example, if you are working with industry you must ensure that the company does not keep commercial secrets about their future direction or market competition that could influence output. This could happen if a company did not want to publish results at a given time due to concerns about competitors using the results for their own commercial gain, for example. It could also occur if a company wanted to delay results until the market is ready for their new product or service. Advice about avoiding such problems is given in Chapter 21. If you are considering collaborating with industry see Appendix 4 for a list of useful websites.

LEAD AGENCY AGREEMENT: COLLABORATION BETWEEN THE US AND UK

A new agreement reached by Research Councils UK and the US National Science Foundation, called the Lead Agency Agreement, came into effect in September 2013. This agreement removes some of the barriers facing international collaboration and simplifies the process for UK and US researchers, enabling you to apply for collaborative funding with minimum extra paperwork. More information about this agreement can be obtained from the websites of Research Councils UK (www.rcuk.ac.uk) and the National Science Foundation (www.nsf.gov).

SUMMARY

Collaboration projects help to share and transfer skills and knowledge, generate knowledge and ideas, increase research output and raise the profile of researchers and their work. They are carried out for political, economic, scientific, professional and social reasons. Although, in most cases, the benefits to be gained outweigh the possible disadvantages, it is vital that you choose your collaborators with care and ensure that all issues of finance and ownership are addressed from the outset.

When you decide to collaborate on your research project you will need to provide a sound justification for the collaboration when you apply for funding. This issue, along with other aspects of budget justification, is discussed in Chapter 11.

Chapter 11

Justifying Your Budget

Justifying your budget is an essential part of the costing process. Funding bodies will not provide funds that are inappropriate or cannot be justified in relation to the proposed research. The justification process explains how the costs were estimated and justifies the need for each cost. It is referred to as a budget narrative or budget justification.

You will need to build your budget narrative, highlight, explain and provide a rationale for each section (or line item) and show that your research will provide good value for money. Also, you will need to make it clear that any request for funds is appropriate and consistent with your university, employer and/or the funding organization policy.

BUILDING YOUR BUDGET NARRATIVE

Large funding organizations explain how to structure your narrative and provide advice about what should be included. Smaller funding bodies may leave decisions about structure, style and content to individual applicants. In general, you will need to include the following categories when building your narrative:

- **Personnel** A detailed list of, and justification for, each member of staff that will work on the project (see Chapter 9). You will need to demonstrate that staffing levels are sufficient and appropriate to the needs of the project.

- **Fringe/employee benefits** An explanation of, and justification for, the benefits package, which can include holiday entitlement, sick leave and health insurance (see Chapter 9).

- **Staff development** For example training, seminars or workshops. These will need to be justified in terms of their importance and relevance to the research project.

- **Travel** A description of, and justification for, the travel expenses of project personnel. You should quote the cheapest prices (at economy level, for example). Taxi fares, accommodation, tips and parking may need to be included.

- **Equipment** Items that need to be purchased (or rented) for the success of the project. You will need to include procurement methods and a justification for using a particular model/supplier.

- **Supplies** A description of, and justification for, items such as office supplies, phone and internet services. All items should be listed separately with costs based on current prices.

- **Consumables** Items such as laptops, software and charges for access to research data. Each item should be listed separately with costs based on current prices.

- **Professional services/consultants** A detailed list of costs (fees and expenses) associated with outside professional services or consultants. The use of professional services and the procurement methods used will need to be explained and justified.

- **Data preservation, data sharing and dissemination costs** Detailed information about costs associated with sharing and making public the results of your research. Careful justification of these costs will help you to produce your impact statement (see Chapter 15).

- **Other costs** Costs that are not listed in the above categories (often because they are unique to a particular project). These will need to be listed and justified in relation to your research.

- **Exceptional items** For example equipment costs over a specified amount, studentships and survey costs. You must be able to demonstrate the importance of exceptional items and justify their inclusion in relation to the success of your project.

- **Indirect costs** For example library facilities and estates, listed and justified if a funding body agrees to pay all or a proportion of these costs (see Chapter 8).

TIP: BE SPECIFIC

Your budget narrative should follow funding body instructions as closely as possible, providing as much detail and justification as necessary, while working within any page length or word count limits. Be as specific as possible. Write your narrative in the same order as the budget line items so that reviewers can compare them easily. Your budget narrative should answer questions rather than generate new ones.

HIGHLIGHTING, EXPLAINING AND PROVIDING A RATIONALE

Most funding bodies will expect you to produce an explanation and rationale for each section of your budget, illustrating how you have arrived at each cost estimate. In general, there are three ways to describe the basis for a cost: actual cost (e.g. salaries); vendor price lists or quotations (e.g. equipment or train tickets); and prior experience (e.g. supplies and services).

When providing a rationale for each of these costs, consider the following questions:

- Who is to be employed on the project? What is their experience, how will they contribute to the success of the project and does this justify the proposed costs?

- If you are using standard job titles and salary levels, does the required work differ in any way from that required on other projects by staff with the same job titles and salaries?

- Can your proposed charges be related accurately to your project? For example, if members of staff are dividing time or resources between projects, how can you ensure that your costs are accurate?

- Does your proposed charge differ from other projects? If so, how does it differ from the standard level expected on all research projects? What is the explanation for this difference?

- How, exactly, will supplies and equipment be used on the project? How are they relevant to your research methods?

- Have you highlighted the availability and status of similar equipment and the anticipated extent of use? If you are collaborating with others, have you included all equipment owned by all parties?

- What are your reasons for choosing a particular model or service contract, in relation to alternatives?

- If upgrades to existing equipment are required, why is this so and why does this provide the best value for money (see below)?

- If it is only possible to provide estimates for certain costs, can you demonstrate that your estimates are sound (based on previous experience or detailed knowledge, for example)?

Sample budget narratives (or examples of best practice) can be downloaded from the websites of many university research offices and provide good examples of how you should work through this process. Some research offices also provide templates that you can use to generate your budget justification for certain types of funding (such as federal funding in the US). Most of this information is freely available and can be accessed by all researchers, even if you are not a researcher in academia.

TIP: THE FIVE 'W's

Consider the five 'W's when justifying costs: who, what, when, where, why? For example, when justifying staff costs, ask the following questions: who are the members of staff? What is their role on the research project and what skills and experience can they bring to that role? When and for how long will they be working on the project? Where will they be working? Why are they required for the project?

PROVING VALUE FOR MONEY

All funding bodies will want to know that all funds they provide will be used wisely and that the results of the research will be worthwhile. For example, in the UK under the Financial Memorandum between HE institutions and HEFCE, and the HEFCE Audit Code of Practice, all institutions have a clear responsibility to obtain 'value for money'. HEFCE uses the term to assess whether or not an organization has obtained the maximum benefit from the goods and services it both acquires and provides, within the resources available. Value for money is discussed in terms of the three 'E's – economy, efficiency and effectiveness – and these provide a useful way for you to work out whether your research provides value for money:

- **Economy** Are you using resources in the best way possible? For example, how will requested equipment save time and effort? Is it possible to show how you will do more for less money?

- **Efficiency** Will your research be carried out in the most efficient way? For example, do members of the team have the necessary skills and experience to carry out the required work with minimal disruption? Do you have examples of good practice that encourage efficiency?

- **Effectiveness** Is your research going to be effective? What is the intended impact and benefit to society? Can you demonstrate how your study will provide a good return on the investment?

TIP: BE REALISTIC

All costs must be realistic, reasonable, justifiable, allowable and allocable (necessary for the success of the project) and your budget narrative must display this clearly. Ask an experienced colleague to check that this is the case.

ENSURING FUNDS ARE APPROPRIATE

Chapters 8 and 9 highlighted the importance of understanding costing methods, caps and limitations. When justifying your budget, you must ensure (and illustrate) that your figures are appropriate in terms of both funding body policy and your organization's policy. Funding bodies will not provide more than their stated cap, however persuasive your argument. They will also want to see that all requests are consistent with your university/employer policies in terms of salary levels, benefits and so on (in most cases your budget will need to receive internal verification that it is appropriate, complete and accurate before you submit: see Chapter 12).

You will also need to ensure that your figures are appropriate for your research project. It is easy for funding bodies to see when budgets have been transferred from one project to another with little attention paid to specific project costs. This will lead to an unsuccessful grant application. In particular, all direct costs can only be charged if they can be identified readily and specifically with your proposed research (see Chapter 8).

SUMMARY

Budget justification is one of the most important parts of your grant application. You must be able to demonstrate that you are only applying for appropriate funds, that all funds are realistic, specific to the proposed research and necessary for the success of your project. You must also be able to prove value for money in terms of economy, efficiency and effectiveness. All sections of your budget justification must adhere to funding body and institutional policy.

Once you are completely satisfied that you have covered all aspects of your budget justification, you can go on to finalize your budget.

Chapter 12

Finalizing Your Budget

Once your research costs have been produced and justified, you can finalize your budget. It is important to get everything right at this stage because once you submit your budget you cannot change or add information. Incomplete forms can lead to budget rejection.

Finalizing your budget involves careful presentation in the exact format required by the funding body and submitting your budget for approval from your institution (it may require approval from your finance department and your human resource department, for example).

PRESENTING YOUR BUDGET

When presenting your budget, take note of the following issues:

- Become familiar with the method required for submission and ensure that your budget and justification are presented appropriately. Many funding bodies require electronic submission that follows a set format. Guidelines are provided, but if you are confused about the system contact the funding body for clarification. See Chapter 16 for more information.

- Ask an experienced colleague, your employer or research office to check your budget and budget justification. Make sure that everything is correct: funding bodies may allow early-career researchers to resubmit if there are minor errors, but this is often not the case for experienced researchers. Most electronic systems enable you to create a PDF of your application that can be printed for checking purposes.

- Make sure that you have enough time to present your budget for institutional approval and the collection of all relevant signatures before the research call/project deadlines (see below). If you are responding to a call for research, ensure that you read the announcement in detail so that you can present your budget in the appropriate way at the appropriate time. This is of particular importance if you are collaborating with others (see Chapter 10).

See Part 3 for further information about producing, presenting and submitting your grant application (including your budget) to funding bodies.

TIP: BE AWARE OF NATIONAL TAX RULES

Ensure that you have taken account of all national tax rules when presenting your budget. For example, in Australia the Goods and Services Tax (GST) applies to almost all research (exceptions include scholarships and exported research). GST is a tax of 10 per cent on all goods and services in Australia and on goods imported into the country (it does not apply to goods and services purchased overseas). In the UK, however, research services are considered outside the scope of VAT. Your finance office will be able to offer further advice.

RECEIVING APPROVAL

In most cases your budget will need to receive approval (or verification that it is current, complete and accurate) before you submit your grant application. This approval could come from your finance office, treasurer's department, research director, research officer or employer. In some cases you may also need to seek approval regarding salaries and associated costs (such as validation of grade roles) from your human resource (HR) department. This is of particular importance if HR is to take on the responsibility of recruitment and selection for your project. All approval must be asked for, and received, well before submission deadlines. Universities and other specialist research organizations will not allow you to submit a grant application until all the required approval has been given.

Also, once you have submitted your complete grant application, some funding bodies will require approval direct from your administrative authority (the organization that will receive the financial payments from the funding body if your grant application, is successful). Once you have submitted your application the funding body will notify your administrative authority automatically, if further approval is required. More information about this procedure is provided in Chapter 16.

TIP: SUBMIT A DRAFT BUDGET

If you need to obtain approval for your budget from your finance office, you can save time by submitting a draft budget before you prepare your case for support (see Section 3). You can continue writing your grant application while the finance office works on your budget, making improvements and/or suggestions for changes (where appropriate) before approval is given.

BUDGET INFORMATION CHECKLIST

The following checklist will help you to determine whether you have all the required information to finalize your budget. If you answer 'no' to any of these questions re-read the relevant chapter in this book or seek further advice from your research office, finance office or an experienced colleague.

	Yes	No
Do you need permission to apply for funding and, if so, has it been granted?		
Have you contacted the research office/finance office for help with your budget?		
Do you understand, and have you adhered to, funding body costing methods?		
Do you understand, and have you adhered to, the required budget format?		
Does your budget fall within the maximum funding level?		
Have you consulted with HR about staff costs and salaries?		
Have you produced a full and accurate list of all required resources?		
Are all direct costs listed, accurate and justified?		
Are all indirect costs listed, accurate and justified?		
Do you know what costs are allowable and are these included in the budget?		
Do you know what costs are not allowed and are these excluded from the budget?		
Do you know, and have you adhered to, the stipend rate for graduate students?		
Do you know, and have you adhered to, caps on salary?		
Do you know, and have you adhered to, institutional fringe benefits rates?		

	Yes	No
Do you know, and have you adhered to, Facilities and Administrative (F&A) costs rates?		
Have you provided a clear justification for every section of the budget?		
Have all collaborators seen and approved of the budget?		
Have you completed all relevant approval forms?		
Have you obtained all relevant signatures?		
Have you received all necessary approval to submit your application?		
Have you got time to complete the rest of the grant application before the deadline?		

TIP: MATCH BUDGET TO PROPOSAL

'We welcome budgets that are detailed and that reflect the research proposal – including details around incentive payments, support to community groups or service users who will be involved as partners or participants, any costs relating to language or access requirements of research participants.'

Joseph Rowntree Foundation (www.jrf.org.uk; Twitter: @jrf_uk)

SUMMARY

When finalizing your budget you must ensure that it is presented in the specific format required by your chosen funding body. You must provide all requested information, ensuring that your figures are correct and that all sections include a justification for costs. Some organizations request that you seek approval before you submit your application. This could be from the finance department or human resource department, for example. Many funding bodies will also require administrative authority approval once you have submitted your budget.

Producing and submitting a budget is only one part of the grant application process. You will also need to produce and submit a project plan, ensuring that you adhere to all funding body policy and guidelines, and meet stated deadlines. These issues are discussed in Part 3.

PART 3
MAKING APPLICATIONS

Chapter 13

Seeking Grant Application Advice

Whereas Part 2 covered costing projects and producing a budget for your research, Part 3 goes on to consider the application process, demonstrating how to produce a good project plan and make a successful application.

If you are an early-career researcher or new to the grant application process you must seek as much advice and guidance as possible in order to increase your chances of success. Competition for grants is fierce and it can be very difficult to obtain funds, especially if you don't work for a high-profile, research-intensive university. There are, however, organizations and departments that have been set up to help novice or early-career researchers.

USING UNIVERSITY RESEARCH OFFICES

University research offices (or research administration offices) manage and support research activity across all disciplines within universities. These offices will undertake most or all of the following:

- manage pre- and post-award activity related to funding;
- offer advice about finding grants and dissemination of funding information;
- provide guidance and assistance with the development of research proposals;

- offer training in applying for grants;

- offer advice about university authorization for submission (see Chapter 16);

- provide advice on contractual matters;

- negotiate contract terms with funders and collaboration agreements with other universities and public sector collaborators (see Chapter 10);

- encourage networking and bring together interdisciplinary groups of researchers with common interests;

- provide advice and guidance on research ethics and governance (see Part 4);

- provide advice about conflict of interest (see Chapter 20);

- develop and implement research strategy and policy;

- manage and monitor research performance;

- administer research budgets;

- co-ordinate activities in preparation for the Research Excellence Framework (REF) in the UK and for similar systems in other countries.

If you work in a university and are thinking about making a grant application, contact your research office for advice. If you are a researcher in industry and are thinking about a collaboration project with a university, you can also make use of the university research office. This also applies to self-employed or retired researchers who have maintained close ties with a particular university. Most research office websites can be accessed by all researchers (whether or not you work in academia) and provide useful advice and guidance about the grant application process.

USING PUBLIC BODIES AND GOVERNMENT DEPARTMENTS

Chapters 1 and 2 listed some of the important public bodies and government departments that provide research funding. These organizations also provide comprehensive information and advice about making grant applications. This information is available on websites and some of the organizations will, in addition, offer tailor-made advice that is specific to your needs. Visit the website of the relevant organization listed below and see Chapters 1 and 2 for more public body and government websites:

Academic researchers

- Australian Research Council (AUS) www.arc.gov.au

- Brazilian National Council for Scientific and Technological Development (BR) www.cnpq.br

- Estonian Research Council (EE) www.etag.ee

- European Research Council (EU) erc.europa.eu/funding-and-grants

- Indian Council of Social Science Research (IN) www.icssr.org

- National Centre for Scientific Research (FR) www.cnrs.fr

- National Research Council Canada (CAN) www.nrc-cnrc.gc.ca_

- National Research Council of Thailand (TH) www.nrct.go.th

- National Scientific Research Fund (BE) www.frs-fnrs.be

- National Scientific and Technical Research Council (AR) www.conicet.gov.ar

- Research Councils (UK) www.rcuk.ac.uk

- Research Foundation Flanders (BE) www.fwo.be

- Slovenian Research Agency (SI) www.arrs.gov.si

- Spanish National Research Council (ESP) www.csic.es_

- Swiss National Science Foundation (CH) www.snf.ch

Health/medical researchers

- Canadian Institutes of Health Research (CAN) www.cihr-irsc.gc.ca

- Health Research Council of New Zealand (NZ) www.hrc.govt.nz

- Medical Research Council of South Africa (SA) www.mrc.ac.za

- National Health and Medical Research Council (AUS) www.nhmrc.gov.au

- National Institute for Health Research (UK) www.nihr.ac.uk

- National Institutes of Health (US) www.nih.gov

Industry researchers

- Enterprise Connect (AUS) www.enterpriseconnect.gov.au

- Industrial Research Assistance Program (CAN) www.nrc-cnrc.gc.ca

- Technology Strategy Board (UK) www.innovateuk.org

TIP: GET EXPERT ADVICE

The National Institute for Health Research (NIHR) has a Research Design Service that 'supports researchers to develop and design high-quality research proposals for submission to NIHR and other national, peer-reviewed funding competitions for applied health or social care research'. The national NIHR Research Design Service has ten regional bases in England. Visit the NIHR website for more information: www.nihr.ac.uk/research.

USING RESEARCH ASSOCIATIONS

There are various research associations throughout the world that support the work of researchers, and which offer information and advice about making grant applications. For example, in the UK Vitae (www.vitae.ac.uk) is a national organization 'championing the personal, professional and career development of doctoral researchers and research staff in higher education institutions and research institutes'. It is funded through the Research Careers and Diversity Unit of Research Councils UK and managed by the Career Development Organisation. You can find information about applying for funding and managing your research career on its website. You can also find information about the UK Research Staff Association.

Other research associations that may be of interest include the following (these sites also provide information about collaboration projects):

- Asia Pacific Educational Research Association (Asia) www.aperahk.org

- Association of Asian Social Science Research Councils (Asia) www.aassrec.org

- Canadian Association of Postdoctoral Scholars (CAN) www.caps-acsp.ca

- Global Research Council (worldwide) www.globalresearchco uncil.org

- International Consortium of Research Staff Associations (worldwide) icorsa.org

- International Council for Science (worldwide) www.icsu.org

- International Social Science Council (worldwide) www.worldsocialscience.org

- International Society of African Scientists (worldwide) theisas.com

- Irish Research Staff Association (IRE) www.irsa.ie

- National Association of Science and Technology Researchers (POR) anict.pt

- National Postdoctoral Association (US) www.nationalpostdoc.org

- Science Europe (EU) www.scienceeurope.org

WORKING WITH EXPERIENCED COLLEAGUES

All new and early-career researchers should try to work with an experienced colleague: your mentor, line manager, supervisor or a senior member of your research team. This will help you to gain valuable experience in making grant applications, and also provides the opportunity for useful feedback on your research proposal and budget. When choosing a colleague to work with, consider the following issues:

- Does your colleague have plenty of experience in obtaining research grants?

- Do you have a good working relationship with your colleague, or the potential to develop one?

- Is your colleague interested and committed to helping you?

- Does your colleague have the time available to help?

- Is your colleague likely to give you a full and honest appraisal?

- Can you handle a full and honest appraisal?

Academic sponsoring

Academic sponsoring is becoming more popular in universities around the world. An experienced researcher agrees to act as a sponsor for an early-career researcher, vouching for them by putting them forward for a particular opportunity. This could be recommending them for a new research team or acting as a referee and mentor for a grant application, for example. There are several benefits to be gained from this type of agreement: for the early-career researcher it enhances career opportunities and increases the likelihood of winning a grant; for the sponsor it displays research leadership and intellectual mentorship; for the research team/department it increases research output and raises their research profile.

If you are a novice or early-career researcher in academia, a sponsor could benefit your career and help you to win grants. As a self-employed, industry or retired researcher, an academic sponsor can help you to access grants that would otherwise be unavailable. You can find a sponsor by getting to know the experts in your field, using university research offices, building networks, attending conferences and using social media (see Chapter 2).

OBTAINING ADVICE FROM FUNDING BODIES

All funding bodies provide information and advice about making a grant application. Some funding bodies are very specific, providing their own templates with comprehensive guidance notes. All this information is freely available from websites (see Chapters 2, 3, 4, 5 and 6 for funding body websites, or use the databases listed in Appendix 2).

Most funding bodies also encourage you to contact them for further advice. They receive far more applications than they can fund, so many are happy to reduce their workload by advising you as to whether your application is relevant and has a chance of success prior to you making your application. Visit funding body websites to find relevant contact details. More information about the importance of establishing a good working relationship with a funding organization is provided in Chapter 17.

SUMMARY

If you are new to the grant application process it is essential that you seek advice before you make an application. Advice is available from university research offices, public bodies and government departments, research associations, experienced colleagues and the funding bodies. Listen to all feedback and take appropriate action so that you can make the best possible grant application.

In addition to understanding how to make a good grant application, it is important to know more about the application process, as this can differ between funding bodies and, in some cases, can be quite complex. These issues are discussed in Chapter 14.

Chapter 14

Planning Your Grant Application

Now that you have gathered information and advice about applying for funding, you can start to plan your grant application. To do this you need to find out about application deadlines, register your details with appropriate bodies and seek all the necessary approvals, gather together your team and choose referees.

SETTING DEADLINES

When setting grant application deadlines you need to consider both external and internal deadlines. Where a funding body has set a specific deadline (often in terms of time and date) you must ensure that all information is collected and ready for submission well in advance of that date. This protects against unforeseen problems such as computer crashes, referees on holiday or problems with gaining internal approval.

Different types of internal approval are required, depending on your organization, the type of research and the grant for which you are applying (see over). Research offices, for example, may require you to complete a summary form that outlines your research proposal, and finance offices may need to approve your budget. These approvals will need to be sought before you submit your application, so find out how long this internal approval process takes and make sure that you can obtain all approvals before the submission deadline.

TIP: KEEP AN EYE ON DEADLINES

Internal deadlines can be affected by time of year, so take this into account when submitting work for approval. University research offices, for example, receive many more grant proposals in the summer months, which can slow down the approval process.

REGISTERING AND OBTAINING APPROVALS

As the applicant it is your responsibility to register and obtain approvals. Funding bodies understand that it may not be possible to receive all approvals before you submit your application, but they will not release funds until all they have all been received. You must take this into consideration when determining the start date of your work. Registration and approvals that might be required (either before or after submission) depend on your research and include:

- **Head of school/employer approval** In most cases this will be needed before you apply for funding. They will need to approve your grant application and demonstrate that you will be supported and employed throughout the length of your project.

- **Research Ethics Committee/Institutional Review Board approval** This is required for all research that involves human subjects: see Chapter 22.

- **Research grant application approval** Universities require a summary of your research to be sent to the research office for authorization prior to grant submission. Researchers working for other research organizations may need to obtain similar approval from the relevant department.

- **Finance office or treasurer's department approval** Often this is needed to verify that your budget is current, complete and accurate (see Chapter 12).

- **Medicines and Healthcare Products Regulations Agency (MHRA) Clinical Trial Authorization (CTA) approval** This is required in the UK to ensure that products are safe and effective, and that the quality of the product is sufficient. More information about CTA approval can be obtained from the MHRA website: www.mhra.gov.uk. Similar approval will be required in other countries.

- **Insurance or indemnity arrangements for negligent or non-negligent harm** These must be in place before the research commences and must clearly define and allocate research roles. Research Ethics Committees are responsible for ensuring the adequacy of these arrangements (see Chapter 22).

- **Sponsorship agreements** These are required for researchers who need a sponsor for their research and grant application. A sponsor is an experienced researcher who is willing and able to vouch for another researcher (see Chapter 13). This differs from a Sponsored Research Agreement (STA), which is an agreement between a funding body/sponsor and a researcher, once funding is granted (see Chapter 17).

- **Data protection registration** In the UK the Data Protection Act 1998 requires every data controller (e.g. organization or researcher) who is processing personal information to register with the Information Commissioner's Officer (ICO), unless they are exempt (see Chapter 22). More information can be obtained from the ICO website: www.ico.org.uk. Similar registration may be required in other countries.

- **Honorary contracts** With the appropriate organization, such as NHS Trusts in the UK for health or medical research.

- **Trust R&D approval for each site at which the research will be conducted** In the UK your health/medical research cannot proceed without formal approval from the R&D office(s) at each of the NHS Trusts in which your research is to take place. Similar approval will be required in other countries.

GATHERING YOUR TEAM

All team members – such as principal investigator, co-investigator(s), collaborator(s) and consultants – should be gathered together in the planning stages so that you can work together on the preparation of your grant application. Most projects will require an experienced principal investigator who is able to demonstrate substantial experience in the field through publication and grant awards. However, some grants are aimed specifically at early-career researchers. If this is the case you should recruit experienced members to your team, or ensure that your sponsor has the required expertise (this could be a supervisor or experienced colleague: see Chapter 13).

The composition of your team will depend on your field of research. For example, a large-scale medical research project may require a clinical expert, an intervention expert, a statistician, a qualitative expert, a health economist, a health psychologist, an organizational psychologist, patient and public representatives and any other required medical specialists. A small-scale social science project, on the other hand, may only require one researcher with experience in the proposed methodology and skilled in the proposed methods.

Choosing team members

Research teams can be built using a top-down approach (by research leaders and experts in their field) or a bottom-up approach (by early-career and novice researchers at the grass-roots level). Both are effective if members of the team are chosen carefully. Team members can be found within your department/organization, through networking opportunities (social media, conferences or workshops), through professional associations, from publications or though recommendations from experienced colleagues.

When choosing team members, take note of the following:

- Cross-disciplinary teams from diverse backgrounds and sectors can promote mutual learning and increase innovation. However, all team members must be willing to foster cultural, social and organizational understanding (see Chapter 10). Check that this is the case at the interview stage.

- When choosing members for the team find out what experience they have working in a research team (successes, failures and challenges).

- Discuss the goals of the research at interview stage. Ensure that members are willing to prioritize the goals of the project, rather than their own objectives.

- Discuss roles, responsibilities and expectations. Members must be willing to work for the team, rather than for their own personal gain.

- Communication is extremely important: team members should have good communication skills and be willing to discuss and share research ideas and data.

TIP: DEALING WITH CONFLICT

Disagreement and conflict can occur among members of a research team, especially in the early stages. Be prepared to discuss and sort out problems early, so that they do not escalate. A team member skilled in conflict resolution and negotiation can be a useful addition to your team.

Checking CVs

Once you have found potential team members, all CVs should be checked carefully. Grant applications will require information about the education, employment history and academic responsibilities of the principal investigator, co-investigators, named postdoctoral students and named research students (see Chapter 15). You must ensure that all are as good as possible and relevant to your proposed research. CVs and justification for team member salaries will need to be included with your grant application.

CHOOSING REFEREES

Most grant applications will require references from scholars who can comment knowledgeably on your research proposal. Referees should know enough about the field to make a meaningful contribution to your application. You will need to contact potential referees prior to submitting your application to check that they are happy to be nominated. Make sure that referees are able to see your research proposal well in advance of producing a reference and, if required, discuss your proposal with them.

Once you have made your submission you are responsible for ensuring that all references reach the funding organization by the stated deadline (see Chapter 16). You can make this process easier for your referees by providing stamped addressed envelopes (or instructions about electronic submission) and ensuring that they are very clear about deadlines.

SUMMARY

When planning your grant application it is important to find out about all internal and external deadlines. Approval and authorization may need to be sought from various internal offices, such as research offices and finance offices and, if this is the case, you need to obtain all the required information before you can submit your application. It is also important to gather your team early in the planning process so that you can work on your grant application together. Referees should be chosen early so that you have time to discuss your plans with them and ensure that they will be able to complete the reference by the submission deadline.

The next stage in the process is writing your grant application.

Chapter 15

Writing your Grant Application

Once you have gathered together all the information required for your grant application you can complete the application form. (For advice on making a successful grant application, see Chapter 17.) To do this you need to be aware of the type of information required, understand each separate component of the form and know what attachments are needed.

In addition to your budget and budget justification (discussed in Part 2), other important parts of the application are your case for support, impact statement and data management plan.

UNDERSTANDING GRANT APPLICATION FORMS

Grant application forms vary depending on the funding body to which you are applying, your subject area and your country. In general, however, you will be expected to complete most or all of the following components (the order and length of each of these will vary):

- **Description/abstract** This should be clear and succinct, and can be used to help the funding body choose the most appropriate reviewer. It might also be published by the funding body, so take care not to use commercially sensitive or confidential information (see Chapter 16).

- **Title** This should be short and explanatory and can hint at your research question, methodology and research population.

- **Subject** You will need to classify your proposal in terms of subject(s) and keywords. This information is used by the funding body to help select reviewers/panels for your grant application. It can also be used for referencing purposes.

- **Academic beneficiaries** You will need to provide information about how other researchers in your field will benefit from your research (a short summary only: further information about communication and dissemination will be included in your 'case for support': see page 126).

- **Institute/organization details** Some funding bodies require additional information about your organization and its ability to manage researchers and oversee research projects.

- **Non-academic partners** This section asks that you list all non-academic collaborators. These must be qualifying organizations and, in most cases, can be national or international (some funding bodies place restrictions on international collaboration, so check that your partners are eligible). See Chapter 10 for more information about working with collaborators.

- **Impact summary** This requires you to answer two questions: who benefits from your research, and how do they benefit? This does not include other researchers, but could be individuals, groups, specific organizations or the wider public. In the UK research councils might publish this information so it should not include commercially sensitive information (see Chapter 16).

- **Attachments** You will be required to provide all the necessary attachments. This could include:

  - *Curriculum Vitae* Basic information about the education, employment history and academic responsibilities of the principal investigator, co-investigators, named postdoctoral researchers and named research students.

- *Publications list* A list of publications by the principal investigator, co-investigators and named researchers. Publications that are of relevance to the research proposal should be highlighted.

- *Visual evidence* Can be provided in support of the proposed aims and objectives or research methods.

- *Technical appendix* Includes relevant technical information that supports the application. In the UK, researchers applying for research council funding may need to produce a Technical Summary and Plan that describes planned technological and digital output.

- *Financial resources and justification of resources* An extremely important component of the grant application process (see Chapter 11). This is where you describe, explain and justify all costs.

- *Impact statement* Expands on the impact summary described above. In the UK, researchers applying for research council funding will need to include a Pathways to Impact attachment that describes the potential impact of your research beyond academia (see below).

- *Work plan/research plan* Provides detail of the work required and the role and responsibilities of those undertaking the work.

- *Head of department/employer statement* This is required to support your application. It demonstrates that your institution intends to support you throughout your project and that your contract will be in place for the duration. It can also contain information in support of your project and in support of you, as the best person to undertake the research.

- **Case for support (or project plan)** This is an extremely important part of your grant application and is described in detail below.

- **Data management plan** Again, this is an important part of your grant application and is described in detail below.

TIP: PAY ATTENTION TO CONTENT

'For research projects, we find that many proposers spend too much time setting out the background to their proposal, and too little time setting out their proposed methodology and analytical framework, and relevance to policy and practice.'

Joseph Rowntree Foundation (www.jrf.org.uk; Twitter: @jrf_uk)

PRODUCING A CASE FOR SUPPORT

A case for support, or project plan, is an extremely important part of your grant application. Again, different funding bodies require slightly different formats for this part of the application but, as a general guide, you will need to include some or all of the following components:

- **Research question or problem** A clear, concise and complex question around which your research is focused. It needs to be specific and defendable, well formulated, credible and easy to understand. It should lead to research that extends or adds to existing knowledge.

- **Aims and objectives** The aim is the overall driving force of your research: a simple and broad statement of intent that describes exactly what you want to achieve. It should emphasize what is to be accomplished and address the outcomes of your project. The objectives are the means by which you intend to achieve the aim. They are detailed and more specific statements that describe specifically how you are going to address your research question.

- **Research background, significance or context** A rationale for your research. Why are you undertaking the project? Why is the research needed? This discussion should be placed within the context of existing research and/or within your own experience or observation.

- **Research design, methodology and methods** A description of your proposed research methodology and a justification for its use. You need to illustrate how your methods (research tools) relate to your methodology and demonstrate how these tools are the most appropriate to answer your research question. Why have you decided to use these particular methods? Why are other methods not appropriate? Do you envisage any problems and how do you expect to overcome them? This section needs to include details about samples, numbers of people to be contacted, methods of data collection and analysis, and ethical considerations.

- **Project management** Detail about who will manage the project and will demonstrate competence and experience.

- **Dissemination** This section demonstrates how you intend to let others know about the results of your research and should be written in conjunction with your impact summary and impact statement.

- **Statement of eligibility** This is required by certain funding organizations and provides evidence that you are eligible to apply for the particular type of funding that is being offered.

See Appendix 4 for more information about producing a research proposal.

WRITING AN IMPACT STATEMENT

Funding bodies will require an impact statement that clearly demonstrates the societal and economic benefits to be gained from your research. It should demonstrate how your research will benefit people, communities, industry or the environment. Your impact statement should describe the issue or problem, provide an action statement, describe the potential benefits and provide a list of researchers, collaborators and contributors along with their impact activities.

TIP: REMEMBER YOUR AUDIENCE

When writing an impact statement, keep your audience in mind and pitch your statement accordingly. Audience members can include funding body panels, external reviewers, peers, government, industry representatives and alumni.

Pathways to Impact

If you are applying for research council funding in the UK, you are required to complete an impact summary (see above) and a Pathways to Impact attachment. This expands on the two questions answered in your impact summary by addressing the question: 'What will be done to ensure that potential beneficiaries have the opportunity to engage with this research?' Your Pathways to Impact attachment will need to include:

- Types of impact activities
    - potential exploitation (commercial and non-commercial);
    - the shaping of policy and practice;
    - application of intellectual assets and outputs;
    - communication activities (e.g. workshops, publications, websites, media relations);

- public engagement activities (past, present and future);

- collaboration relationships, roles and responsibilities in terms of impact activities;

- members of the research team, involvement, roles and responsibilities in terms of impact activities.

- Impact activity deliverables and milestones

  - timescales for delivering impact activities;

  - key milestones for impact activities;

  - methods used to measure success of impact activities (i.e. monitoring and evaluation methods).

- **Summary of resources for the impact activity** (the bulk of this resource listing and justification will be in the Financial Resources and Justification of Resources sections of the main application form).

TIP: AVOID USING WAFFLE

Make sure that your impact statement is clear, concise and specific. Don't waffle or pad your statement with irrelevant material. When requesting resources for impact activities you must be able to demonstrate that all activities are project-specific and justified.

PREPARING A DATA MANAGEMENT PLAN

Most funding bodies require the preparation of a Data Management and Sharing Plan or a Data Management Plan (DMP) either as part of the main proposal or to be attached to grant applications as a supplementary document. The plan is reviewed as an integral part of the grant application: if it is required and you fail to include it in your submission, your application will be returned without review.

A DMP is used to describe what type of data will be generated and how. It sets out how the data will be stored, preserved, shared and disseminated. It also includes information about security and possible restrictions, given the nature of the data. Producing a DMP is useful because it helps you to justify the resources and funding requested and helps to clarify researcher and institutional roles and responsibilities.

If such a plan is requested, the funding body will have specific data management policy to which you should adhere. You will also need to adhere to your own institutional data management policies. Although policy varies between funding bodies and organizations, it will include some or all of the following issues:

- All data are expected to be of the highest quality and should have long-term validity.

- Researchers are expected to share with other researchers all primary data, samples and supporting materials (in a form that protects individuals and research participants). Exceptions can be granted if a legitimate case is presented.

- Data should be well stored and preserved in an accessible form so that other researchers can access, understand, use and add value to the data.

- All data should be stored, preserved and managed with the highest regard for ethical standards.

- All data should be stored, preserved and managed within the law, regulation and recognized good practice.

- Researchers should share, or make widely available, any inventions or software developed from the research. However, they will be able to retain principal legal rights to intellectual property.

- Researchers are expected to prepare and submit for publication within a reasonable time.

- Authorship of publications should be claimed only by those who have been involved in the work and should reflect the amount of their contribution towards the work.

- Funding bodies will encourage publication (unless the funding body itself elects to publish or disseminate findings).

Many funding bodies produce guidance notes and templates for producing a DMP, and these are available on websites or electronic submission systems. Further information and advice about curating and preserving research data can be obtained from the Digital Curation Centre (DCC) (see Appendix 4). The DCC website contains a useful online tool that can be used to create a DMP.

SUMMARY

Although grant application forms vary, depending on your chosen funding body and the country in which you are applying for funding, there are some components that are required in all applications. This can include an abstract, project title, case for support/ project plan, impact statement and data management plan. Pay close attention to application instructions and guidelines: failure to include required components will lead to application rejection.

Once you have completed your grant application and ensured that you have included all the requested information, you can go on to submit your application form.

Chapter 16

Submitting your Grant Application

Once you have produced your project plan and finalized your budget you are ready to submit your grant application. This chapter provides information about the submission process, including producing a covering letter, receiving administrative authority approval, choosing external reviewers and submitting commercially sensitive material. It concludes with a comprehensive submission checklist.

WRITING A COVERING LETTER

If the submission process allows it, send a covering letter with your grant application. This should include the title of your proposed research, a very brief description of your project, information about the call for research/type of grant for which you are applying and concise information about yourself, with contact details.

Other information can be included if relevant. For example, some funding bodies point out that it would be helpful to receive the names of people who you would not wish to act as reviewers because of conflict of interest (see page 134). Other funding bodies, however, have a specific section on the application form for this information.

TIP: PAY ATTENTION TO DETAIL

When making your submission, pay close attention to detail. Most funders will reject applications that have not correctly addressed issues such as page length, font size and supporting evidence (such as CVs, quotations, justification of resources, impact statements and letters of collaboration). Funding organizations, in most cases, will not allow you to resubmit this information.

RECEIVING ADMINISTRATIVE AUTHORITY APPROVAL

In Chapter 14 we saw that some funding bodies require you to inform the administrative authority of your host institution of your intention to apply for funding prior to making your submission. (The administrative authority is the institution that will receive the financial payments if your application is successful.) In these cases you can only submit your application once such approval has been received.

In addition to this, once you have submitted your application it will be sent to your administrative authority for further review and approval. This procedure has been simplified by the electronic submission process. For example, in the UK the Research Councils (BBSRC, EPSRC, ESRC, MRC, STFC and NERC) and the Technology Strategy Board (TSB) use the grant application portal Je-S (je-s.rcuk.ac.uk). This electronic system operates a two-stage submission and approval process whereby you, as the principal investigator, submit an online application. Your university research office is notified automatically of your submission and asked to authorize your application on behalf of the administrative authority (e.g. your university).

In most cases this is a simple process because you have already submitted your application for review prior to making your grant submission (see Chapter 14). However, research offices may be unaware of funding body deadlines, so you should inform them of these to ensure that all approval has been given before the deadline. A similar system is adopted by The Royal Society (e-gap.royalsociety.org) and the British Academy (egap.britac.ac.uk).

More information about the submission and approval procedure can be obtained from your chosen funding body, your research office or research manager/supervisor.

TIP: CHECK YOUR ATTACHMENTS

Electronic submission systems have internal validation processes (to check word counts in dialogue boxes, for example), but this does not extend to attachments. When attaching additional information (covering letters, CVs, letters from collaborators, and so on) you must ensure, personally, that all requirements in terms of style, structure and word count are met.

CHOOSING EXTERNAL REVIEWERS

Funding bodies may request that you make suggestions for external peer reviewers for your grant application. Although rules vary between funding bodies, you should take note of the following when suggesting external reviewers:

- Reviewers must not be collaborators, colleagues or researchers with whom you have published work recently. Some funding bodies give a specific period of time: the Wellcome Trust (www.wellcome.ac.uk), for example, specifies five years.

- It may be possible to name researchers who you don't want to see your research proposal. However, you cannot name all researchers in your field and you will have to provide justification as to why a particular researcher should not be consulted.

- If you decide to nominate a reviewer you should not contact that person, except to ask their permission to be nominated. Nor must you discuss your work with any nominated person.

- Funding bodies may choose not to use any of your nominated reviewers, and instead choose other reviewers to assess your application.

TIPS FOR SUCCESSFUL SUBMISSION

- If you are asked to suggest external reviewers, choose wisely.
- Meet the application deadline, ensuring that all necessary approval meets this deadline.
- If supporting evidence is to be sent separately, make this clear in the covering letter. Follow up to ensure all supporting evidence has arrived safely.
- If references are to be submitted separately, follow up to ensure that they have been sent.
- Provide all documentation requested, in the requested format.
- Use the specified font type and size.
- Do not exceed page limits.

SUBMITTING COMMERCIALLY SENSITIVE MATERIAL

When you submit your application you will be required to include an abstract of your research. This will be used to draw attention to your research and could be placed on funding body websites or in directories. Therefore, you must take care not to use commercially sensitive or confidential information in your publishable abstract. Also, research councils in the UK require you to submit an impact summary (see Chapter 15). It is possible that this information is published so avoid commercially sensitive material in this section of your grant application form.

If you have concerns about sharing commercially sensitive data that is generated from your research, it is possible to establish a reasonable binding agreement that contains confidentiality provisions. Most funding bodies want you to share your data with other researchers, but they are also sensitive to the issues of intellectual property and potential commercial value. Discuss this issue with your funding body prior to making your application as this information will be required for your data management plan (see Chapter 15).

COMPLETING A SUBMISSION CHECKLIST

When you feel that your application is ready for submission, work your way through the following list of questions. If you answer 'no' to any of these you may need to undertake further work before submitting your application.

	Yes	No
Have you made a clear case for why the research is needed?		
Have you demonstrated the likely impact of your research?		
Have you identified the research problem clearly?		
Have you defined your research question(s) clearly?		
Do you have well-defined aims and objectives?		
Have you summarized existing evidence and/or observation?		
Have you shown how your research will add to this evidence and/or observation?		
Have you shown that there is a gap in existing knowledge?		

	Yes	No
Have you shown how your research will fill that gap in knowledge?		
Have you shown how your proposal meets the requirements set out in the brief?		
Have you highlighted existing relevant data sources?		
Have you provided a clear statement about your study design?		
Have you shown how your proposed methodology will answer your research question?		
Have you demonstrated that your methodology will meet your aims and objectives?		
Have you justified your methodology and methods?		
Have you provided a rationale for sample size?		
Have you produced a realistic data collection plan?		
Have you supplied a detailed data management plan, if requested?		
Have you demonstrated that your timetable is realistic and feasible?		
Are all your figures accurate and correct?		
Have you provided a rationale and justification for all budget items?		
Have you defined all key terms?		
Can your proposal be understood by experts and lay people?		

	Yes	No
Have you avoided jargon and used plain English?		
Have you demonstrated that suitable project management/leadership is in place?		
Have you given a clear description of your research team?		
Does your research team include all relevant expertise?		
Have you demonstrated how your research provides good value for money?		
Have you obtained all required approvals?		
Have you attached all the necessary paperwork?		
Have you provided suitable references and ensured they will be completed on time?		

TIP: DON'T WORK IN PARALLEL

Don't be tempted to submit parallel applications. In most cases, larger funding bodies will not consider applications that are the same as one that is currently under consideration (by the same funding body or by another). Funding bodies may share pertinent information to make sure that you have not made a parallel application.

SUMMARY

When submitting your grant application you must ensure that you have included all the required information and that it is sent in the requested format. Many funding bodies will return applications that are incomplete or in the wrong format and, in some cases, you will not be able to resubmit (there may be some leniency for new or early-career researchers). If the submission format allows it, include a covering letter with your application and take advantage of the offer to choose external reviewers, if available. Take care when submitting commercially sensitive material and ensure that this is not included in your abstract or impact statement.

There are other steps you can take to improve your chances of success when applying for grants, and these are discussed in Chapter 17.

Chapter 17

Making Successful Grant Applications

Competition for research funding is fierce so it is important to do all that you can to increase your chances of success. This involves careful justification of your research topic, question and methodology; ensuring compatibility between you, your research and the funding body; building relationships with the funding body; proving your ability, knowledge and expertise; and producing the best possible application. Once you have done this and have been successful in obtaining your grant you must understand, and adhere to, all contracts and funding agreements.

JUSTIFYING YOUR RESEARCH

Successful applicants are able to justify their research in terms of topic, research question and methodology/methods. The advantages and disadvantages of the project must be weighed carefully and you will need to demonstrate the strategic importance and impact of your research. This helps reviewers to understand the relevance, timeliness, importance and uniqueness of your project. In particular, you will need to pay close attention to justifying the following:

- **Topic** Working through a process of topic justification will help reviewers to decide whether your proposed project is viable, workable, exciting, important and worth funding. It is also important to consider how you decided on the topic: is it a valid, reliable and justifiable reason for undertaking the research?

- **Research question** This guides you through the research and writing process. Your research project will succeed only if you have asked the right question. It is important, therefore, to be able to justify your question and demonstrate to reviewers that you have asked the right one. This question must also have relevance to the objectives of the funding organization (see below).

- **Methodology** Methodological defence is an extremely important part of your grant application, especially for those who choose a more innovative, less well-known methodology. You must be able to demonstrate that your methodological choice fits with your theoretical perspective and epistemological standpoint, and that your methodology is the most suitable framework for your chosen topic and research question. Take care to avoid methodological fundamentalism when defending your choice (this implies that your methodology is the one true approach and that all others are inferior and/or flawed). Be prepared to critique your methodology and change/adapt/combine if necessary.

TIP: BE HONEST

'We welcome proposers who are explicit about the limitations and risks of their proposed approach, not just the merits. It gives us confidence where proposers have clearly thought hard about what could go wrong (beyond everyday project management issues).'

'We welcome proposals that give an honest and thoughtful account of what aspects of diversity and equality they will incorporate in their work, and how they will do this (rather than simply stating "we will sample for..." but without then referring to how ethnicity, gender, age, sexuality etc. will feature in the analysis and approach).'

Joseph Rowntree Foundation (www.jrf.org.uk; Twitter: @jrf_uk)

IMPROVING YOUR CHANCES OF SUCCESS

You can improve your chances of success if you take note of the following:

- **Build relationships** Researchers who win funding tend to be good at establishing relationships with funders, are proactive in forming partnerships, understand which funders match their needs and interests and adhere to all funder guidelines, seeking advice where appropriate. It is important for funders to build a relationship and collaborate with researchers over the duration of the project. If you fight against this collaboration you are less likely to be funded.

- **Ensure compatibility** Funding bodies tend not to work with researchers whose focus and approach are not compatible with theirs. Ensure that all applications are aimed specifically at your chosen funding body.

- **Ensure that your application is pitched at the right intellectual level** and demonstrates the following:

  - you know about, and can critically analyze, current thinking in your discipline;

  - you will create and interpret new knowledge (through research and advanced scholarship);

  - your work will advance the knowledge in your field;

  - you are able to design and implement a project that will enable you to generate new knowledge;

  - you have the background understanding and required knowledge of relevant research techniques (including an evaluation and critique of methodology and methods);

  - you are able to critique, justify and modify your project in light of problems that can arise;

- you will be able to publish your work and it will satisfy peer review.

- **Publish as much as you can** In most cases you can only cite published or accepted-for-publication papers when making a funding application. Although some funding bodies target early-career researchers, in most cases your publication record will have an influence on funding decisions.

- **Produce the best possible application** Edit, re-read, proof-read and proof-read again. Ask experienced colleagues, your employer and staff in your research office to read your application. Try to obtain objective views. Make alterations, based on experienced feedback. See Chapter 18 for information about avoiding common application mistakes.

- **Provide evidence of good management/leadership** Successful applicants are able to demonstrate that they have a competent, experienced person who can lead the project and ensure that it meets its aims and objectives. They are also able to demonstrate that there will be good financial management and that there are efficient procedures in place for the handling of finances, including good bookkeeping and properly prepared accounts.

TIP: LEARN HOW THE FUNDING PROCESS WORKS

Sitting on internal or external funding panels, or becoming an external reviewer, is a useful way to find out how the funding process works and will enable you to understand more about how funding is granted. Large funding bodies produce comprehensive guidelines for reviewers and these can help to improve your own grant applications.

RECEIVING YOUR GRANT

Once you have successfully received your funding, in most cases you will be required to submit a copy of the feedback letter or email from the funding body/sponsor to your employer, administrative authority or research office. You will be given a funding agreement or contract that outlines the responsibilities of all parties (the chief/principal investigator, the administering organization and the funding body). Although agreements vary, they tend to include the following:

- when funds will be paid;
- how funds will be paid;
- how the funds can be used;
- specified personnel;
- transfers;
- audit and monitoring;
- reporting requirements;
- termination of grant.

Where variation of contract is permitted as the project progresses, you will also be given details of how to request a variation and information about the procedures that must be followed. Your university research office, manager or employer will be able to offer further advice about these and other contractual issues.

TIP: STUDY SUCCESS RATES

Research councils, government departments and many trusts and charities publish information about recent awards and success rates for grant applications. Study the statistics: get to know more about the funding body, who they fund, how much they provide and how successful researchers are when they make an application. This will help you to choose a funding body that provides the greatest chance of success.

SUMMARY

Researchers who are successful in their grant application are able to demonstrate the importance, timeliness and relevance of their research. Their proposed research is compatible with the purpose and ethos of the funding body, and they are able to demonstrate a willingness to develop a good working relationship throughout the duration of the project. Researchers who have a good publication record (or a willingness to publish and disseminate results effectively if they are new or early-career researchers) are more likely to succeed in their application.

However, even the best researchers can have their grant application rejected. Coping with rejection is discussed in Chapter 18.

Chapter 18

Coping with Unsuccessful Grant Applications

Unfortunately, many grant applications are unsuccessful. This can be extremely disappointing and disheartening, especially when you have put in considerable time and effort to your grant application. Although the main reason for failure is that there is not enough funding for all applicants, there are various others reasons that can lead to an unsuccessful application.

Rejection, however, does not signify the end of your project: you may be able to resubmit your application after minor alterations, or you can approach another funding body.

REASONS FOR APPLICATION FAILURE

There are many reasons why a grant application might fail. These include the following:

- the applicant and/or project do not meet the eligibility criteria;

- the application form and/or additional paperwork are not submitted within the deadline;

- submission guidelines have not been followed and/or applicants have misunderstood or not complied with funding regulations;

- application forms are incomplete and requested items of information are missing;

- the proposal is poorly prepared and lacks professionalism (poor writing, poor punctuation, poor grammar and spelling mistakes);

- there is a mismatch between the aims of the research and the aims of the funding body;

- aims and objectives are unclear or incomplete;

- there is a lack of focus or aims and objectives are too general;

- researchers do not have the required experience and/or CVs are not up-to-date;

- there is no evidence of adequate project and financial management;

- figures are incorrect, missing or naïve;

- costs have not been justified adequately;

- the conceptual framework is badly developed;

- the methodology is flawed (not suited to the research question, for example);

- methodological detail and information about research methods is lacking;

- the significance and intended impact of the project have not been adequately explained;

- the research is more appropriately funded by another organization, either public or private;

- the proposal is similar to projects that are already funded by the organization.

Even if you believe that you have addressed all these issues and have presented the best possible proposal, competition for funding is fierce. All funding bodies receive many more applications than they can possibly support. It takes time, patience, skill and experience to apply successfully for funding.

If you do not succeed don't take it personally. Instead, read all reviewer feedback carefully and, if you have the chance, revise your application and resubmit. If the funding body does not allow resubmission, you can apply to another funding body (see below). More information about reasons for proposal failure can be obtained from *50 Mistakes Grant Writers Make* (see Appendix 4).

TIP: DON'T GIVE UP

One of the best ways to cope with rejection is to have another grant application almost ready for submission. Although you should never submit parallel applications, you can develop several proposals for relevant funding bodies that are ready to submit one at a time (as long as each proposal is specific to the funding body to which you intend to apply). Use the feedback from one rejection to enhance and improve the next application.

RESUBMITTING AN APPLICATION

It may be possible to resubmit your application with minor alterations. First, check that resubmission is allowable under funding body rules. If it is, read the review/summary statement from the reviewers/funding body to ascertain whether the problems are fixable (see box over). Seek advice from experienced colleagues or your research office if in doubt. Also, try to work out whether the reviewers were the right people for the job: are they experts in your field, did they understand your methodology and your rationale? It may be possible to resubmit and assign different reviewers, although you will need to check that this is the case before resubmission.

Resubmitting an application with minor alterations can be a quicker procedure than for the original application as you might not need internal approval from various departments before submission. When resubmitting your application answer all reviewer questions/concerns as specifically as possible. Highlight or draw attention to changes that have been made as a result of reviewer feedback. In most cases you will only be able to make one resubmission, so make sure that it is as good as it possibly can be.

Don't take offence when your application is criticized. Read all reviews carefully and adjust your work accordingly. If you are convinced that you are right and the reviewer is wrong, it is possible to try to persuade them that your view is the correct one, but you must do so in a courteous and professional manner, using reasoned judgement and balanced argument.

Potential fixable problems	Hard-to-fix problems
Poorly presented application form	Unoriginal research
Poor spelling, grammar and punctuation	Topic doesn't match funder objectives
Information not presented in the correct format	Badly developed hypothesis
Required information missing	Research will not add to knowledge
Lack of expertise for certain sections of the project	Lack of expertise from whole team
Not enough justification for the topic	Inappropriate methodology
Insufficient background information	Lack of importance
Insufficient methodological/ methods discussion	Insufficient impact
Insufficient discussion of potential problems	Costs cannot be justified
Lack of references	Poor references
Management/leadership not emphasized	No evidence of research experience
Timetable/timeline is incomplete	Financial instability of organization

Resubmitting a new application to the same funding body

If you choose this option you must ensure that your new submission is substantially different in content and scope (with greater and more significant differences than are normally encountered on resubmitted applications). The new application could include a change in direction and approach, changes to the research plan, changes to your methodological approach, different aims and objectives or a revised research question (and the consequences on potential outputs and impact), for example.

You cannot simply retitle your work, change your abstract and ask for different reviewers. Funding bodies will screen your application to check that it is not a resubmission of the same or similar project. For example, the National Institutes of Health in the US screens in the following way:

'New applications received by the NIH are screened multiple times and checked to determine if the application is a new application. The first check is done within the Division of Receipt and Referral in the Center for Scientific Review (CSR). Subsequent checks are performed by the Scientific Review Officer in charge of the review meeting and by NIH program staff.

'Sometimes reviewers identify potential problems. Previous applications are analysed for similarities to the current application and Summary Statements are also considered. When an application is determined to be another version of an application that has already received the maximum number of reviews and in violation of this policy, the "virtual A2" application is administratively withdrawn and is not processed further. The Division of Receipt and Referral, CSR, informs the project director/principal investigator and institution of this determination.'

National Institutes of Health (grants.nih.gov/grants/policy)

APPLYING TO ANOTHER FUNDING BODY

If your application to one funding body has failed you can apply to another. If you decide to do this, take note of the following:

- Submission requirements between funding bodies can vary widely. Make sure that you obtain the most up-to-date guidelines from the new funding body and adapt your submission accordingly.

- Available funds, caps and limitations vary between funding bodies. Again, you must obtain the most up-to-date guidelines and ensure that each section of your budget falls within these caps and funding limits.

- Reviewers can tell when a proposal has been adapted only slightly from one funding body to another. This can lead to application failure, so you must make significant changes to your proposal (or even re-write) where necessary (see the Tip below).

TIP: MATCH PROJECT WITH FUNDER

'We welcome proposals that really fit the brief and guidance we've issued (it is generally easy to identify the projects that have been put forward "on spec" but were probably developed for another funder).'
Joseph Rowntree Foundation (www.jrf.org.uk; Twitter: @jrf_uk)

SUMMARY

Grant application rejection can be hard to take. You have worked extremely hard on your application, you are an expert in the field, your research is important, timely and relevant and the public will gain considerably from your output. However, reviewers and funding bodies seem to think otherwise. It is important not to take their comments or rejections personally. Listen to all feedback and alter and resubmit your application, if you get the opportunity. If not, find another funding body to which you can make an application.

Part 3 has offered advice about making an application for funding, illustrating how to increase your chances of success and knowing how to cope with rejection. Part 4 goes on to consider ethical issues, which includes seeking ethical approval for research that involves human subjects, avoiding conflict of interest and avoiding biased financial relationships.

PART 4
ACTING ETHICALLY

Chapter 19

Addressing Ethical Issues

All researchers should act with integrity, paying close attention to ethical issues. Reputable funding organizations will request information about ethical issues in their funding application and all research involving human subjects will need ethical approval before grants are given and studies can commence. Other researchers too, when reviewing and critiquing your work, will want to see that these issues have been addressed.

MEETING ETHICAL CRITERIA OF FUNDERS

You must act ethically when dealing with funding bodies. Research is more likely to be supported and funded if researchers are seen to be accountable and act with integrity. Also, reputable funding organizations will only fund research that meets their ethical criteria. These will be displayed on websites, sent with grant application forms or made available on request. When submitting a grant application you will need to demonstrate that you are aware of, and have addressed, all the ethical implications associated with your research.

When applying for funding you should only use reputable funding organizations that act ethically. Your university or employer will have strict rules about the origins of research funding: your research project and source of funding will need to be approved by the relevant research ethics committee (or approval board) before your studies can commence (see Chapter 22).

TIP: THINK ABOUT ETHICAL ISSUES

'A common weakness of proposals is to not address ethical issues fully. This can, on occasion, result in an otherwise strong proposal failing to be funded if a competing proposal has given sufficient information or, at best, funding being delayed as we clarify this issue with the applicant.'

Joseph Rowntree Foundation (www.jrf.org.uk; Twitter: @jrf_uk)

ACTING ETHICALLY WITHIN RESEARCH AIMS

You must act ethically within the general research aims of discovering knowledge and avoiding error. This means acting with integrity and honesty and not fabricating, falsifying or misrepresenting data. It is possible for some researchers to face pressure from employers or sponsors to produce positive results and ignore negative results, for example. As an ethical researcher you must resist this pressure and act with integrity. Chapter 20 offers advice about avoiding conflict of interest and Chapter 21 covers avoiding biased financial relationships.

TIP: MAINTAINING STANDARDS AND INTEGRITY

A 'concordat' about standards and integrity in UK research has been developed by Universities UK in collaboration with research councils, the Wellcome Trust and various government departments. This 'sets out five commitments that assure Government, the wider public and the international community, that the highest standards of rigour and integrity will continue to underpin research in the UK'. A copy of the document can be downloaded (or a hard copy ordered) from www.universitiesuk.ac.uk.

ACTING ETHICALLY WITHIN MORAL AND SOCIAL VALUES

You must act ethically within established moral and social values. This includes the following:

- respecting human dignity, privacy and fundamental rights;

- paying close regard to animal welfare (visit the website for the National Centre for the Replacement, Refinement and Reduction of Animals in Research for information: www.nc3rs.org.uk);

- taking note of, and adhering to, issues of health and safety for all members of the research team and for research participants (see tip, below);

- avoiding discrimination.

TIP: HEALTH AND SAFETY IN RESEARCH

A document called *Responsible Research: Managing Health and Safety in Research* can be downloaded from the Institution of Occupational Safety and Health website: www.iosh.co.uk. It has been produced in association with the Universities and Colleges Employers Association and the Universities Safety and Health Association and provides a useful guide on health and safety issues in research. Although this document has been produced in the UK, it is relevant to researchers worldwide.

ACTING ETHICALLY WITH PARTICIPANTS

You must treat research participants with dignity, privacy and autonomy, minimize harm and risk and consider issues of informed consent and data protection. All participants should be treated with respect and nobody should be exploited, bullied or cajoled into taking part in research or offering an opinion.

Taking part in research can affect people in different ways. Some may find participation a rewarding process, whereas others will not. Your research should not give rise to false hopes or cause unnecessary anxiety. You must try to minimize the disruption to people's lives and if someone has found it an upsetting experience you should find out why and try to ensure that the same situation does not occur again.

Some of the people who take part in your research may be vulnerable because of their age, social status or position of powerlessness. If participants are young you need to make sure that a parent or guardian is present. If participants are ill or reaching old age you might need to use a proxy, and care must be taken to ensure that you do not affect the relationship between the proxy and the participant. All these issues will need to be addressed in your funding application.

Anonymity

You must do your best to ensure anonymity by showing that you are taking steps to make sure that what participants have said cannot be traced back to them when you disseminate your results.

You need to demonstrate in your data management plan how you intend to categorize and store information and show how you are going to make sure that the information is not easily accessible to third parties. Also, participants need to know that what they say cannot be used against them in the future. Reputable funding organizations will require this information when you make your application. More information about producing a data management plan is provided in Chapter 15.

Confidentiality

To ensure confidentiality you need to show that information supplied to you in confidence will not be disclosed directly to third parties. If the information is obtained in a group setting, issues of confidentiality should be relevant to the whole group (each member will need to agree not to disclose information directly to third parties).

You will need to demonstrate how you are going to categorize and store the information so that it is kept safe and secure. Again, reputable funding organizations will require this information in your data management plan. See Chapter 22 for more information about the 'duty of confidentiality' that exists in UK law and Chapter 15 for more advice about your data management plan.

ACTING ETHICALLY WITH THE PUBLIC

When dealing with the wider public you must strive to promote social good through your research practice and results, and take action to mitigate social harms through your research.

Issues of trust are high on the political and public agenda at the present time. Therefore, your research must be seen to be trustworthy and you should strive to ensure that your work is respected by the wider public. Conflict of interest and biased financial relationships must be avoided as these reduce levels of trust and can have a detrimental impact on your research and on research practice in general (see Chapters 20 and 21).

COLLABORATING AND CO-OPERATING ETHICALLY

You must act ethically when you collaborate and co-operate with other researchers/academics. You must be trustworthy and accountable and act with mutual respect and fairness. Close attention must be paid to the following issues:

- copyright;
- patents;
- data protection (see Chapter 22);
- plagiarism;
- intellectual property.

> ### *TIP: BUILDING TRUST*
>
> You can build trust and respect by communicating openly and honestly, face-to-face, where possible. Include all colleagues in decision-making and respect others' opinions, strengths and expertise. Be reliable: always complete work to deadlines. If things go wrong, don't blame others, but work together to achieve a solution to the problem. This is of particular importance when working with collaborators (see Chapter 10).

PUBLISHING AND DISSEMINATING RESULTS ETHICALLY

You must act ethically when publishing and disseminating results. This involves issues of confidentiality and anonymity, in addition to respecting the wishes and rights of those who have taken part in the research.

It is important to resist all sponsor/funder control of, or influence on, publication decisions. When applying for funding, try to ensure that publication decisions and editorial control remain with you, as the researcher. Don't be bullied or cajoled into publishing positive and suppressing negative results, for example, or delaying publication until your sponsor feels the time is right, commercially (see Chapter 21).

If your research involves the generation of large-scale datasets, reputable funding organizations will require you to submit details of your data management and sharing plan and will request that their funded researchers maximize the availability of their data with as few restrictions as possible (see Chapter 15). Small-scale projects will not need this type of plan, but funders will request that data is made available to other researchers on publication and that data is deposited appropriately and in a timely manner.

SUMMARY

It is important for researchers to address all ethical aspects of their research during the planning stages. Funding bodies and sponsors will want to see that this has been done and you will need to present this information to ethical review bodies when you submit your grant application for review. It is important that research is trusted and respected by the wider public, politicians, research participants and other researchers, and this can be achieved through careful analysis, documentation and reporting of ethical issues.

An important ethical issue that you will need to address is conflict of interest. This is of particular importance for researchers who are sponsored by private organizations or who have previous connections with their sponsor.

Chapter 20

Addressing Conflict of Interest

Chapter 19 highlighted some of the important ethical issues that are pertinent to all researchers who are applying for funding. Another ethical issue that is of great importance is conflict of interest. This can arise when one or more researchers on a project have additional interests, which may corrupt their studies by undermining their impartiality in some way. It can also occur where a funding organization or employer has interests that may affect the progress or outcome of a research project.

RECOGNIZING CONFLICT OF INTEREST

Various types of conflict of interest can arise during a research project. This type of conflict occurs when outside influences affect the impartiality of your research. These could be the personal, political or financial interests of the researcher(s), the financial or reputational influence of the employer or the political influence of a particular funding body. The following (real) examples illustrate the types of conflict of interest that can occur.

Example 1

A self-employed researcher is commissioned by the principal of a local college to undertake some research into students' experiences of their courses. The college is to go through an inspection and the researcher is told that the results of the research must reflect well on the college. Indeed, it is made clear that payment for the research is dependent on a positive outcome.

Example 2

A pharmaceutical company requires all their researchers to sign a contract that prevents them from providing information about their employer, their research or their results to members of the public or media. A researcher is suspicious of some results concerning a new drug, but has been told by his employer that the results are sound and must be replicated, 'using any means necessary'.

Example 3

A trustee of a neighbourhood community group is married to a local authority employee who finds out that funds are available to support the group's work. However, the group must provide evidence to prove that the local community benefits from the service. The trustee is passionate about her community group and decides to undertake the research herself so that she can prove that the group serves the local community well.

Example 4

A senior researcher at a red-brick university sits on the advisory panel of a national research council. He surprises his colleagues by putting in a bid for funding in a subject outside his area of expertise and on a theme that is not, normally, funded by the research council. His bid is successful. When the research council publishes its themes for the next round of funding, this particular theme is included in the list.

Example 5

In a clinical trial one team member is responsible for patient selection. However, that team member owns stocks in the company that is supporting the trial (see Chapter 21 for more information about this type of financial conflict of interest).

These examples illustrate that conflict of interest can occur at various stages of the research process: it can influence the topic of the research, the design of the project, the methodology, the reporting of results or it can be the reason for the research being conducted in the first place. As an ethical researcher, it is important that you avoid conflict of interest at all stages of your work.

PRESSURE FROM GOVERNMENT– COMMISSIONED RESEARCH

Researchers from the London School of Economics found evidence to suggest that governments lean on academics to produce findings that chime with political objectives. Researchers in the study reported that civil servants interfered throughout the different stages of the research process. For example, some would insist on changing the case-study sample so that specific groups that agreed with government policy were chosen to offer their opinions, and others would make it clear from the outset that the purpose of the study was to prove that a new programme was cost-effective. However, in most cases researchers were able to stand up to this interference and political pressure did not influence their conclusions. The report of the study can be read online: the LSE GV314 Group, 'Evaluation under Contract: Government pressure and the production of policy research', *Public Administration* (published online October 2013: onlinelibrary.wiley.com).

AVOIDING CONFLICT OF INTEREST

Various techniques can be employed to avoid conflict of interest:

- Learn how to recognize and identify when conflict of interest occurs.

- Find out all you can about the funding body, your employer and other researchers on your team.

- Understand why the research has been commissioned and be clear about the justification for the research.

- Seek out information so that you can comply with current rules and regulations. Large funding bodies will have detailed guidelines and policy about conflict of interest and many of these are updated regularly. Also, universities offer training related to the regulations and their institution's financial conflict of interest policy.

- If your organization does not have a written policy about avoiding conflict of interest, produce your own guidelines. These should include the following issues:

  - Your research must not be influenced (or seen to be influenced) by any outside factors. This can include personal, political or financial interests.

  - The potential for personal gain must not jeopardize (or appear to jeopardize) the integrity of the research. This must be the case for all stages of your research, including the choice of research topic, the design of the research, the interpretation and publication of results.

  - You must be open and honest about all financial and commercial interests related to your research. This includes declaring information about the funding body and any other organization that may provide finance or equipment.

- Break all ties with the person or organization that is causing the conflict of interest.

There may, however, be occasions where it is not possible to avoid conflict of interest. If this is the case you will need to take steps to manage and minimize conflict (see below).

TIP: WATCH OUT FOR BIAS

Take care to recognize unintentional bias that can occur as a result of conflict of interest. For example, you may choose a topic and methods for which financial support is available because your research cannot proceed without funding. Or you may choose a more marketable topic because it is easier to publish your results so that you can increase your publication record and enhance your research career. These decisions may be unintentional, but they do have an influence on your research.

DECLARING CONFLICT OF INTEREST

Once identified, all actual or potential conflict of interest must be declared to relevant parties so that they can be made aware of the nature and extent of the conflict. Parties that should be notified include employers, research ethics committees and research participants. Conflicts should be declared as soon as they arise. Some conflicts will only need a written declaration to be kept in your organization/university records, whereas others will need a detailed action plan about how they can be managed (see below).

When declaring conflicts the following issues will need to be considered:

- the type of potential conflict;
- the nature of the conflict;
- a description of all parties involved;
- a description of potential financial interests and rewards (if relevant);
- possible violations of legal requirements (if relevant);
- a description of potential commitment and loyalty interests (if relevant);
- the potential influence of interests on research integrity.

Your declaration will be evaluated by the relevant parties, based on rules, regulations and legal requirements. These parties could decide that no further action is required, or they could ask that a conflict of interest management plan is set up (see below). In some cases they could refer your declaration to a higher review body.

TIP: DISCLOSURE REQUIREMENTS

Some funding bodies set specific disclosure requirements related to conflict of interest. You need to be aware of, and comply with, these requirements. More information will be found in the funding application terms and conditions.

MANAGING CONFLICT OF INTEREST

In cases where it is not possible to avoid conflict of interest, these conflicts must be managed carefully. Review boards or senior managers will want to see that a careful conflict of interest management plan is put into place. This can involve:

- keeping a register of interests;
- complying with all legal requirements and institutional regulations;
- ensuring that the person or organization with the conflict of interest is isolated from the decision-making process for every step of the research process;
- transferring the person with the conflict of interest to another role, or reassigning tasks;
- keeping written records about all decisions made;
- providing a regular update to the review body/senior managers.

SUMMARY

Conflict of interest occurs when a researcher, team member, funding body or organization has conflicting interests that affect the integrity of the research. These conflicts can be financial or political, or relate to commitment or loyalty, for example. They can also occur at various stages of the research process, from choosing a research topic to disseminating results. Conflicts of interest are best avoided to maintain research integrity. However, if this is not possible they need to be declared and carefully managed. Institutional policy is provided to help researchers recognize, declare and manage conflict of interest.

The important issue of inappropriate financial relationships – a major cause of conflict of interest for researchers who are trying to obtain funding – is discussed in the next chapter.

Chapter 21

Avoiding Biased Financial Relationships

Chapter 20 illustrated that a major conflict of interest for researchers is to do with the issue of finance. This is of particular importance to researchers who need to obtain funding for their research. When you seek and obtain funding you must take care to avoid biased financial relationships. These can include funding that comes with strings attached, researchers who own shares in the funding organization, payments and incentives given to members of the research team and funding that originates from unethical practice, for example.

UNDERSTANDING THE FUNDING EFFECT

'Funding effect' is a term that has been coined by social scientists to explain why research outcomes are sometimes significantly different in publically funded and in privately funded research. Initially, this was seen to occur mostly in drug research. However, it is possible for this funding effect to be present in all types of funded research. As an ethical researcher you must take steps to recognize and avoid unethical practice associated with funding effect. This practice can include the following:

- Funding bodies only fund research that is going to portray them or their product favourably (or competitors unfavourably).

- Researchers design their research in way that will attract funding. This includes the type of research question asked, the hypothesis put forward for testing, the methodology chosen, the way data are interpreted and the style of reporting results.

- Researchers are expected to claim superiority of their research, trials or data, without being able to back this up with evidence.

- Researchers are expected to claim superiority of their research, trials or data, when it is not possible to make comparisons because their research has used a different sample, different doses or tested for slightly different outcomes.

- Researchers are pressured to report positive and ignore negative results, thus providing an incomplete scientific record.

- Funding bodies delay reporting outcomes (or suppress results) that could have a negative impact on their commercial activities or reputation.

- Researchers who have a financial stake in the outcome of their research change the way they conduct their research (questions and methods are changed in subtle ways to obtain the desired results, for example).

TIP: WATCH OUT FOR MISCONCEPTIONS

When large sums of money are involved in a research project it can attract misconceptions and rumours about misconduct. You must pay close attention to *perceptions* of research integrity as well as *actual* research integrity.

ACTING AGAINST BIASED FINANCIAL PRACTICE

Maintaining integrity is extremely important for researchers. Yet, in practice, this can be difficult when researchers have to compete for funds, maintain a publication record and remain in their jobs. However, continued funding, public trust and research(er) reputation are dependent on high-quality research that is seen to be objective and free from financial influence.

The following points illustrate how you can choose to act against biased financial practice and maintain research integrity.

- Update your knowledge and/or training in research methods, paying attention to the particular types of bias that can influence how your research is chosen, conducted, analyzed and/or reported.

- Voluntarily refuse to accept funding that is influenced by any of the funding effects described above.

- Steer clear of funding bodies that partake in unethical practices. This will involve detailed research before you fill in a funding application (see below).

- Decide whether your funding body has a significant financial stake in the outcome of the study. If so, approach with caution.

- Find out whether the funding body has a history of influencing research methods or outcomes to promote their financial goals.

- Only use funding bodies with a strict code of ethics that includes issues of financial conflict of interest (see below).

- Check all funding contracts carefully. Ensure that all investigator decisions, publication decisions and editorial control remain with you, as the researcher.

- Try to ensure that your funding contract includes a clause that funders cannot interfere with research methods or results.

- If you have already accepted funding, recognize if and when biased financial practice occurs and take action to acknowledge, reduce or eliminate the problem (see Chapter 20).

- Discuss your concerns with your line manager, employer, supervisor or a colleague.

- Only publish in journals that have strict editorial controls and stringent policies about the sources of research funding. Most reputable journal editors now insist on disclosure of all financial conflict of interest and editors may use information disclosed in conflict of interest and financial interest statements as a basis for editorial decisions.

- Find out whether a funding organization is happy for you to acknowledge the source of funding in all publications. If they do not agree to this, find out why.

- Avoid conducting research for organizations to which you are financially tied (through stocks, shares or equity, for example). If this cannot be avoided all conflict of interest must be disclosed (see Chapter 20) and the necessary ethical approval obtained after disclosure (see Chapter 22).

- Never accept financial rewards, incentives or gifts from funding bodies or companies that will gain from a positive research outcome (this gain could be commercial, financial or reputational).

TIP: AGREE STANDARDS

Problems can be avoided if you agree your research plan and negotiate investigator standards with your funding body before the research begins. Your mutually agreed-upon standards should be included in your contract to promote objectivity, avoid conflict between funding body and researcher and ensure that financial bias is avoided.

REPORTING FINANCIAL MISCONDUCT

If you feel that serious financial misconduct in a research study has taken place you should report it as soon as possible. Advice and guidance specific to your organization will be available in your staff handbook and/or on your organization's website. In most cases you will need to report any suspected misconduct to your line manager, supervisor or governing body. They will have set procedures that must be followed to decide on whether the issue needs to be investigated or reported further, and whether outside bodies need to be brought in to the investigation.

If further investigation is required you will be requested not to discuss any aspects of the investigation, either internally or externally, except during investigation meetings. You will also be told not to speak to the media. When making these requests your employer must not infringe any rights conferred on you by the Public Interest Disclosure Act 1998 (if you work in the UK). This Act covers issues such as protective disclosure, unfair dismissal, national security and compensation. Similar legislation offers protection in other countries. Reputable organizations will also have policy in place that enables staff to report misconduct without fear of reprisal (and protect against harassment and victimization).

Investigation outcomes

The investigation will consider the basis for allegations to ensure that they have not been made through misunderstanding or for trivial or malicious reasons. Various outcomes are possible:

- No further action required.
- Disciplinary action required, in which case the usual disciplinary procedures should be followed.
- Legal advice required and appropriate action taken.

A review of the investigation should be undertaken, with weaknesses identified and action for improvement put in place.

USING ETHICAL FUNDERS

Many of the problems described above can be avoided if you choose your funding organization wisely and ensure that you only use ethical funders. These organizations will have strict codes of conduct and ethics that enable you to conduct your research without influence. Public funding bodies, in particular, pay close attention to these issues because it is a matter of public trust, as described by the National Institutes of Health in the US:

> 'The NIH is committed to preserving the public's trust that the research supported by us is conducted without bias and with the highest scientific and ethical standards. We believe that strengthening the existing regulations on managing financial conflicts of interest is key to assuring the public that NIH and the institutions we support are taking a rigorous approach to managing the essential relationships between the government, federally-funded research institutions, and the private sector.'

National Institutes of Health (grants.nih.gov/grants/policy/coi/).

All research funded by public bodies such as this will need to meet the required ethical standards and you will need to act with integrity, honesty, accuracy and efficiency when conducting this type of funded research. You will also be required to adhere to rules, regulations and guidelines and follow commonly accepted professional codes.

SUMMARY

Researchers have an ethical responsibility to act with integrity and produce the best research possible, free from financial relationships that can undermine the credibility of the research. You should not enter into agreements that interfere with your research methods or the way you publish your results. If you suspect that financial misconduct has influenced a research study you should report the problems as soon as possible. Problems can be avoided if you choose to work with ethical funding organizations that have strict policy about financial relationships.

Chapters 20 and 21 have illustrated how to recognize and avoid conflict of interest that can arise from inappropriate relationships or financial misconduct, for example. The next chapter describes the process of ethical review, which provides further ethical checks for all research involving human subjects.

CHAPTER 22

Applying for Ethical Approval

It is now a requirement in most countries that certain types of research should undergo a process of ethical review. Research cannot proceed until approval has been granted. Although definitions vary between countries, in general, legislation applies to any biomedical and behavioural research that involves human subjects. The purpose of such law is to protect the rights, safety, dignity and well-being of research participants and promote ethical research that is of potential benefit to participants, science and society.

KNOWING WHEN ETHICAL APPROVAL IS REQUIRED

When deciding whether ethical approval is required, you need to determine whether your studies are classed as research. Certain types of study (such as a clinical audit or a service evaluation) are not classed as research and will not need to obtain ethical approval (see over).

The following questions will help you to determine whether your studies are classed as research. If you answer 'yes' to any of these, you will almost certainly require ethical approval. See Chapter 23 for information about ethical review and approval bodies in different countries and to find out more about application procedures and fees (if relevant).

	Yes	No
Are you intending to generate new knowledge with your study?		
Do you intend to generate and test hypotheses?		
Do you intend to identify/explore themes following an established methodology?		
Have you developed a clear question that you wish to answer?		
Have you developed well-defined aims and objectives to answer your question?		
Are you going to collect data that is additional to data collected for routine care?		
Will your studies involve some type of intervention?		
Do you have a sampling frame underpinned by theoretical justification?		
Does your study involve randomization?		

The National Research Ethics Service (NRES) in the United Kingdom has produced a booklet called *Defining Research* that is available for download from the NRES website: www.nres.nhs. uk. Also, the Medical Research Council has produced a Health Research Authority Decision Tool that helps you to assess whether your work is research (see Appendix 4).

KNOWING WHEN APPROVAL IS NOT REQUIRED

Approval should not be required if you are able to answer 'no' to all the questions above. However, knowing exactly when approval is or is not required can be difficult, so if in doubt seek advice from your institutional or local ethics committee (see Appendix 4). The following examples of studies that may not need approval are provided as a general guide, but rules and regulations vary between countries.

- A review of existing literature.

- Research that does not involve human beings or data about them.

- Quality assurance activities or evaluation projects designed for self-improvement or programme evaluation, not intended to contribute to new and generalizable knowledge.

- Clinical audits that are undertaken to provide information that will help to ensure that the best care is given, where existing data is analyzed and there is no allocation of intervention or randomization.

- Service evaluations that are undertaken to judge the service, where existing data is analyzed and there is no allocation of intervention or randomization.

- Usual medical practice where the goal of the activity is to benefit a well-defined group of people in a predictable way, such as blood donations and vaccinations.

- Usual public health practice. This can include monitoring and surveillance of an outbreak or incident using systematic, statistical methods, for example. Data is used to ascertain the source of the outbreak and assess risk, but does not affect treatment.

- The use of public data sets. Researchers must not merge any of the sets in a way that individuals can be identified and must not enhance the public data set with identifiable, or potentially identifiable, data. The use of restricted data sets and those that require the signing of a User Agreement will probably need approval, so seek further advice.

UNDERSTANDING THE ETHICAL REVIEW PROCEDURE

If you decide that your research does require ethical approval, you will need to contact the appropriate person or body. University researchers should contact their supervisor, head of department, university research ethics committee or review board. Postgraduate and undergraduate students should ask their supervisor/tutor for further advice, or visit the website of their university ethics committee, which should contain all the required forms. If your research is considered low risk it may simply need to be certified by your supervisor or by the principal investigator. Medium and high-risk projects will need a higher level of review.

Researchers conducting health and medical studies (including those working within universities) will need to contact the National Research Ethics Service (NRES) or their local research ethics committee in the UK, or the equivalent organization in other countries. The ethical approval process is managed by the Integrated Research Approval System (IRAS) which acts as a single system for applying for the permissions and approvals for health and social care/ community care research in the UK: www.myresearchproject.org. uk. Other researchers will need to obtain details of the most appropriate research ethics committee or review board (see Chapter 23 for more information).

Procedures vary depending on the type of researcher, the subject of the research, the committee or review board and the country in which the research is being undertaken. In general, you will need to obtain and complete a questionnaire or application form and assemble all the necessary documentation, which can include consent forms, participant information sheets, advertisements and questionnaires. You will need to address the legal issues and make sure that you have all the required checks, registrations and insurance in place before you submit your application (see over). Electronic and/or paper applications should be submitted following the relevant guidance.

MAKING AN APPLICATION FOR ETHICAL REVIEW

Most ethics committees and review boards will only accept an application for ethical review after you have secured funding for your research. Most funding organizations accept this rule, but they will still require detailed information about ethical issues on your funding application form. However, if a funder requires you to have ethical approval before they will grant funding, you will need to attach a letter explaining this when you apply for ethical approval.

Although requirements vary (depending on level of risk and your country, for example), in general you will need to address the following questions when applying for ethical review:

- Are you absolutely clear about your research question, aims, objectives and methodology? Are they justifiable in terms of knowledge generation and impact? Can you explain them clearly, succinctly and to a lay audience (committees may consist of experts and lay members: see Chapter 23)?

- Do you need to undergo a Disclosure and Barring Service (DBS) check in the UK or equivalent in other countries (see below)? Have you given yourself enough time to arrange and complete the check?

- Do you need to register for data protection (see page 183)? Have you given yourself enough time?

- Does your research place you or your participants at risk? Have you completed a risk assessment and obtained the appropriate risk forms to complete (see Chapter 19)?

- Are you insured to conduct your research? Universities, hospitals and industry should all have automatic insurance cover for most studies, but you will need to check that this is the case. Also, some insurers may need prior notification of certain types of research. Self-employed researchers and community researchers will need to arrange their own insurance, if required.

- If you are recruiting participants, have you obtained all the necessary recruitment documents? This could include information sheets and adult consent forms.

- Are you required to submit a copy of your questionnaire or interview schedule to the ethics committee or review board? Will you have time to produce this information?

- Are you required to submit a copy of your advertising and recruitment literature, such as posters or leaflets? Again, will you have time to produce this?

- Are you required to submit a copy of your research information sheet and Code of Ethics (that you intend to give to participants)?

- Do you intend to offer incentives or payment? If so, are they justifiable and reasonable? Many organizations have strict rules about how much can be paid to participants and about how and whether payments are advertised.

TIP: GIVE YOURSELF ENOUGH TIME

Remember to give yourself time to go through the ethical review process before your research commences. Research cannot begin until approval has been granted and this can take some time, especially if your application is returned with a request for clarification or referred to a higher level committee (see below).

Ethical review outcomes

The committee will review your application (usually within a stated time) and provide one of the following outcomes:

- accepted;

- returned to you for clarification/additional information;

- referred to a higher level of ethics committee (if your research is medium to high risk and requires a higher level of scrutiny, for example);

- rejected (this is a rare outcome that is used when research is deemed to be ethically or scientifically unsound: you should not resubmit until you have discussed your research with the ethics committee).

PROCEDURES FOLLOWING APPROVAL

Once approval for your research has been given, the principal investigator must advise the ethics committee or review board of any proposed changes or amendments. In most cases you will not be able to implement changes until further approval has been given, unless the changes are required immediately to prevent harm to participants or researcher(s).

You must also report any adverse effects promptly (whether serious or non-serious) and give details of changes required to prevent these occurring again. Many ethics committees and review boards will require an annual report outlining progress (a 'continuing review approval form' may be provided for this purpose).

COMPLYING WITH THE LAW

All types of researcher, including undergraduate and postgraduate students, self-employed researchers, community researchers and industrial researchers, must comply with ethical approval legislation. In addition to a requirement for ethical review, there are other legal requirements to which researchers must adhere, as discussed over. (It is not possible to list all the legislation pertaining to research ethics worldwide. Therefore, contact your university research office, legal team or ethics committee for more information about relevant legislation in your country, if it is not mentioned below.)

Data Protection

The Data Protection Act 1998 is the main piece of legislation that covers the issue of data protection in the UK. In the European Union the Data Protection Directive covers the protection of individuals with regard to the processing of personal data and the free movement of such data. At this present time, the United States does not have comparable, single legislation concerning data protection. Instead, legislation in the US is adopted on an ad hoc basis, combining legislation, regulation and self-regulation.

In the UK the Act relates to all data about living and identifiable individuals that is held, or intended to be held, on computers or in a 'relevant filling system'. This includes contact details, such as telephone numbers, email addresses, names and addresses. It also includes identifiable sensitive data such as health, sex life, criminal record, politics, religion, trade union affiliation, ethnicity and race. However, if your data is anonymous or aggregated (combined from several measurements) it is not regulated by the Act. In these cases extreme care must be taken to ensure that the method you have used to make the data anonymous or aggregated cannot be reversed in any way.

Confidentiality

A 'duty of confidentiality' exists in UK law. Although it can be difficult to interpret this law clearly because it was established through case law rather than by statute, it generally means that confidential or sensitive information should not be disclosed to third parties without prior consent. If a participant supposes that the information supplied is given in confidence, it must be treated as such. You must obtain written consent if you wish to share the data with other academic researchers (under strict terms and conditions). However, there are exceptions, such as when information is subpoenaed by police or courts, or in cases where child abuse is suspected, for example.

Disclosure and Barring Service (DBS)

If your research includes access to children (someone who is under eighteen) or contact with vulnerable adults, in the UK it is now a requirement by law for researchers to undergo a Disclosure and Barring Service (DBS) check (previously a Criminal Records Bureau check). It can take more than a month to arrange and complete the check so you will need to apply well in advance. In most cases you will need to quote your DBS Disclosure Number on your ethical approval application form and some funding organizations will also request this number when you apply for funding.

SUMMARY

All research involving human subjects must be submitted for ethical review. This is a procedure that is carried out by local or national ethics committees or review boards that are made up of a combination of experts and lay members. The purpose of review is to ensure that researchers act ethically and that no harm comes to research participants. It also helps to determine nationally consistent ethical standards in research and provides scrutiny for researchers and their studies.

Chapter 23 lists and describes some of the important ethics committees and review boards around the world.

Using Ethical Approval and Review Bodies

Chapter 22 has illustrated that research involving human participants, human tissue or databases of personal information must be
approved by a recognized research ethics committee or review board
before any research activity can commence. Ethics committees and
review boards have been set up around the world to protect the
rights, safety, health and well-being of human subjects involved in
clinical trials, medical research and any other research that involves
human subjects. The work of approval and review bodies is to
make an assessment of, and ruling upon, the suitability of research
proposals, investigators, research methods and facilities.

This chapter discusses the work of ethical approval and review
bodies (and networks) in the United Kingdom, European Union,
United States, Australia, Canada and New Zealand, and in universities worldwide. Although similar bodies exist in other countries,
it is not possible to list them all here.

HEALTH RESEARCH AUTHORITY (UK)

The Health Research Authority (HRA) was established in 2011
'to protect and promote the interests of patients and the public
in health research, and to streamline the regulation of research'
in the UK. It has taken on the functions of the National Research

Ethics Service (NRES) and this includes ethical review and approval of health research. The HRA establishes NHS Research Ethics Committees within the NRES and these are described below.

The HRA website provides useful information about funding your research, planning and designing your project, determining which review body approvals are required, participation information sheets and informed consent. It also explains more about the Integrated Research Application System (IRAS), an online system for preparing regulatory and governance applications for health and social care research.

NHS RESEARCH ETHICS COMMITTEES (UK)

NHS Research Ethics Committees (RECs) have been established by the HRA to 'safeguard the rights, safety, dignity and well-being of people participating in research in the National Health Service'. The committees consist of between seven and eighteen volunteer members, one third of which must be lay members and the rest experts in their field. Members receive special training in ethical review.

The role of these committees is to review research proposals and give an opinion on participant involvement and whether the research is ethical. The following types of research are reviewed:

- research taking place in more than one domain (for example, a Strategic Health Authority in England, a Health Board in Scotland, a regional office of the NHS Wales Department or the whole of Northern Ireland);

- clinical trials of investigational medicinal products;

- research involving medical devices;

- research involving prisoners;

- research involving adults lacking capacity;

- establishing research tissue banks;

- projects funded by the US Department of Health and Human Services (DHHS);

- establishing research databases.

There are different types of NHS REC across the UK, reviewing different kinds of study. More information about these different types of REC can be obtained from the HRA website. You can also obtain information about finding an REC and booking a review on this site.

EUROPEAN NETWORK OF RESEARCH ETHICS COMMITTEES (EU)

The European Network of Research Ethics Committees (EUREC) 'brings together national Research Ethics Committees (REC), associations, networks or comparable initiatives on the European level'. The network has been established to:

- foster a sustainable infrastructure for European RECs to promote exchange and cooperation and to allow for international cooperation;

- gather information on RECs in Europe to build a basis for mutual exchange;

- collect and evaluate training materials for REC members to enhance the quality of review;

- conduct capacity building to facilitate the development of national REC networks;

- identify emerging ethical issues to develop common solutions for challenges posed by new technologies and scientific methodologies.

You can find a useful interactive map of all REC members in Europe on the EUREC website, which enables you to click on a particular country to obtain European country-specific details of RECs and/or research ethics policy. This is of particular use to you if you intend to undertake a collaborative project with a European partner, for example (see Chapter 10).

INSTITUTIONAL REVIEW BOARDS (US)

In the United States, institutional review boards (IRBs) are committees that have been set up to approve, monitor and review biomedical and behavioural research that involves human subjects. Their role is to protect the rights and welfare of human participants by reviewing research protocols, investigators and related materials. IRB is a generic name and does not have to be used: institutions can choose their own name, which can include 'ethical review board' or 'ethics committee', for example.

IRBs are subject to regulations set by the Food and Drink Administration (FDA) and the Office for Human Research Protections (OHRP), within the Department of Health and Human Services. Specific requirements for each committee are set out by the regulations. This includes a requirement that the committee has at least five members with the necessary expertise, with a minimum of one scientist and one non-scientist. Both men and women must be represented and they should not be all of the same profession. More than half of the members must be present to vote on a proposal and IRB members may not vote on their own proposals.

IRBs are set up by medical facilities or universities that have researchers conducting studies on humans in subjects such as health and the social sciences. There are also commercial, for profit IRBs. If an institution (or an independent or community researcher) does not have an IRB it is possible to arrange for an outside IRB to conduct the review, as long as the arrangement is documented in writing.

You can find an IRB by using the searchable database available on the OHRP website. The database is searchable by IRB name and/ or number, but you can also search your location for a list of IRBs.

AUSTRALIAN HEALTH ETHICS COMMITTEE (AUS)

In Australia it is the role of the Australian Health Ethics Committee (AHEC) to issue human research guidelines. The committee also undertakes a rolling review of the 'National Statement on Ethical Conduct in Human Research' and other projects to address important issues in health and medical research. This guides the work of Human Research Ethics Committees, described below. The committee also provides advice on international developments in health ethics issues.

There are seventeen members of the committee, drawn from areas of expertise such as philosophy, the ethics of medical research, public health and social science research, clinical medical practice and nursing, disability, law, religion and health consumer issues. More information about the AHEC and ethical issues in Australia can be obtained from the National Health and Medical Research Council (NHMRC) website.

HUMAN RESEARCH ETHICS COMMITTEES (AUS)

In Australia the task of reviewing research proposals that involve human beings is undertaken by Human Research Ethics Committees (HRECs). These are committees that are established by public, not-for-profit and private organizations, such as universities and hospitals. Each HREC is encouraged to register with, and report annually to, the NHMRC.

Organizations or independent researchers that do not have an HREC are able to approach established HRECs with a request to review proposals on a regular or ad hoc basis. In these cases some HRECs may charge for their services. You can find a list of registered HRECs (with contact details) on the NHMRC website.

NATIONAL COUNCIL ON ETHICS IN HUMAN RESEARCH (CAN)

In Canada the National Council on Ethics in Human Research (NCEHR) was established in 1989 'to advance the well-being and protection of human participants in research and to foster high ethical standards for the conduct of research involving humans'. It is a non-governmental organization composed of voluntary members who are interested in furthering its mission.

The main role of the NCEHR is to offer assistance and guidance to Research Ethics Boards. These are the committees in Canada (in universities and hospitals, for example) that review applications for ethics approval, oversee and advise on the ethical aspects of research involving human subjects, and provide a resource for education, guidance and leadership on ethical research. More information about the NCEHR can be obtained from their website.

NATIONAL ETHICS ADVISORY COMMITTEE (NZ)

In New Zealand the National Ethics Advisory Committee (NEAC) was established in 2001 to:

- provide advice to the Minister of Health on ethical issues of national significance in respect of any health and disability matter (including research and health services);

- determine nationally consistent ethical standards across the health and disability sector and provide scrutiny for national health research and health services.

The NEAC has twelve members that are appointed by the Minister of Health, although the Committee remains independent of the Ministry and its work. The NEAC issues ethical guidelines for health and disability research that are used by ethics committees to review research proposals. It also sets the standards that must be met by researchers.

The NEAC website contains some useful publications, produced by the Committee, concerning research ethics, along with details of annual reports and reviews of the ethical research process in New Zealand.

UNIVERSITY RESEARCH ETHICS COMMITTEES (WORLDWIDE)

Universities have a panel or committee that is responsible for ethical issues within the university. Tasks and remits vary, depending on the country and type of university, but in general the work of ethics committees (and ethics sub-committees) within universities is to:

- promote an awareness and understanding of ethical issues in research throughout the university;

- offer advice on ethical issues in research;

- keep abreast of external ethical issues in research and ensure that the university responds to all external requirements;

- protect the safety of research subjects and ensure that all research meets the highest academic standards of quality, integrity and ethics;

- undertake reviews of individual research ethics applications, or provide advice and guidance to individual department/ schools about how to undertake their own reviews;

- offer guidance in cases of uncertainty about whether an ethical review is needed;

- reassure researchers that their proposals have taken account of ethical issues and all risk, and that proposals are ready for submission (to funding bodies or higher level ethical review committees, for example);

- hear appeals against decisions that have been made by schools or departments.

As we have seen above, medical or clinical research, in many countries, will require outside approval from a specialist ethics committee or review board. Your university research ethics committee will be able to offer advice about whether this is the case and, if so, provide contact details of the relevant committee or review board (see Chapter 22 for more information about medical and NHS approval).

SUMMARY

Local and national research ethics committees and review boards have been established to scrutinize research that involves human subjects. In most cases, countries have a national committee that oversees and advises the work of local or specialist committees.

This book has provided information about obtaining funding for your research. It includes advice about finding sources of funding, costing your project, putting together a funding application, acting ethically and receiving approval. Obtaining funding for research can be a long and laborious process, but I hope that I have made this process easier by providing all the relevant information in one accessible source. Useful databases, directories and further reading conclude this book and provide useful information for those of you who wish to follow up any of the advice given. I wish you every success in obtaining funding for your research and I hope that your research project proceeds well.

Eligibility for Research Council Funding (UK)

Research Councils UK (RCUK) sets out the eligibility criteria for research funding from the seven research councils in the UK. The following types of organization can apply for research council funding (a full list of criteria can be obtained from the RCUK website: www.rcuk.ac.uk):

- **All UK Higher Education Institutions** The higher education funding councils for England, Wales, Scotland and Northern Ireland determine whether an organization meets the criteria to be a Higher Education Institution.

- **Research Institutes** With which the research councils have established a long-term involvement as major funder (see list of organizations, below).

- **Other Independent Research Organizations (IROs)** May also be eligible if they 'possess an existing in-house capacity to carry out research that materially extends and enhances the national research base and are able to demonstrate an independent capability to undertake and lead research programmes' (see list of organizations, over). They must also satisfy the following conditions:

- Organizations which are, or which are constituent parts of, a charity registered with the Charities Commission; or associations that are eligible for exemption from Corporation Tax.

- The organization must be a legal entity that is not owned, established or primarily (i.e. 50 per cent or more) funded for research purposes by any single part (or related parts) of the public sector (other than by a research council, HEI, NHS Trust, National museum/ gallery/library/archive/botanical garden/observatory) or by a business.

- The organization must possess an existing capability to carry out high-quality research, which requires a minimum of ten researchers with significant publication records, and an excellent track record of research staff.

- Sufficient financial support for research at the organizational level (averaging at least £0.5 million over the previous three years).

- Evidence of the organization having a strong track record of maximizing the wider impact and value of its research to the benefit of the UK economy and society.

RESEARCH COUNCIL INSTITUTES

The current list of eligible Research Council Institutes is as follows:

Biotechnology and Biological Sciences Research Council (BBSRC)

Babraham Institute

Institute of Food Research

John Innes Centre

Rothamsted Research

The Genome Analysis Centre

The Pirbright Institute

Engineering and Physical Sciences Research Council (EPSRC)

European Atomic Energy Community/ Culham Centre for Fusion Energy (EURATOM/CCFE)

Medical Research Council (MRC)

MRC Anatomical Neuropharmacology Unit

MRC Biostatistics Unit

MRC Cancer Cell Unit

MRC Cell Biology Unit

MRC Centre for Protein Engineering

MRC Clinical Sciences Centre

MRC Clinical Trials Unit

MRC Cognition and Brain Sciences Unit

MRC Collaborative Centre for Human Nutrition Research

MRC Dunn Human Nutrition Unit

MRC Epidemiology Resource Centre

MRC Epidemiology Unit

MRC Functional Genetics Unit

MRC Health Services Research Collaboration

MRC Human Genetics Unit

MRC Human Immunology Unit

MRC Immunochemistry Unit

MRC Institute of Hearing Research

MRC Laboratory of Molecular Biology

MRC Mammalian Genetics Unit

MRC Molecular Haematology Unit (including MRC UK Mouse Genome Centre)

MRC National Institute for Medical Research (NIMR)

MRC International Nutrition Group

MRC Prion Unit

MRC Protein Phosphorylation Unit

MRC Radiation and Genome Stability Unit

MRC Social and Public Health Sciences

MRC Toxicology Unit

MRC/Cancer Research UK/BHF Clinical Trial Service Unit

MRC (UK) The Gambia

MRC Unit for Lifelong Health and Ageing

MRC-University of Glasgow Centre for Virus Research

MRC/UVRI Uganda Research Unit on AIDS

The Francis Crick Institute

Natural Environment Research Council (NERC)

National Oceanographic Centre

NERC British Antarctic Survey

NERC British Geological Survey

NERC Centre for Ecology and Hydrology

Plymouth Marine Laboratory

Scottish Association for Marine Sciences

Science and Technology Facilities Council (STFC)

Diamond Light Source Ltd

Isaac Newton Group

Joint Astronomy Centre

STFC – Laboratories

UK Astronomy Technology Centre

INDEPENDENT RESEARCH ORGANIZATIONS

The current list of eligible IROs is as follows:

All NHS Trusts, hospitals, Boards, Primary Care Trusts and GP practices

Animal Health Trust

Armagh Observatory

Ashridge

Beatson Institute for Cancer Research

British Library

British Museum

British Trust for Ornithology

CABI Bioscience UK Centre

Cambridge Crystallographic Data Centre

Cancer Research UK

CERN, the European Organization for Nuclear Research

Chatham House

Christie Hospital NHS Trust

East Malling Research

EMBL – European Bioinformatics Institute

European Southern Observatory

Game and Wildlife Conservation Trust

HR Wallingford Group Ltd

Imanova Limited

Imperial War Museum

Institute for European Environmental Policy

Institute of Development Studies

Institute of Employment Studies

Institute for Fiscal Studies

Institute of Occupational Medicine

International Institute for Environment and Development

Liverpool School of Tropical Medicine

Marine Biological Association

National Archives

National Centre for Social Research

National Gallery

National Institute of Agricultural Botany

National Institute of Economic and Social Research

National Museum Wales

National Museums Liverpool

National Museums of Scotland

National Portrait Gallery

North Bristol NHS Trust

Overseas Development Institute

RAND Europe Cambridge

Royal Botanic Gardens – Edinburgh

Royal Botanic Gardens – Kew

Royal Commission on the Ancient and Historical Monuments of Scotland

Royal Society for the Protection of Birds

Royal United Services Institute for Defence and Security Studies

Sir Alister Hardy Foundation for Ocean Sciences

Science Museum Group

Tate

Tavistock Institute

The National Maritime Museum

The Natural History Museum

The Victoria and Albert Museum

The Young Foundation

Transport Research Laboratory

United National Environment Programme – World Conservation Monitoring Centre (UNEP-WCMC)

Wellcome Trust Sanger Institute

Zoological Society of London, Institute of Zoology

APPLYING FOR IRO STATUS

If you wish to apply for IRO status you should contact the Je-S helpdesk in the first instance (email: JeSHelp@rcuk.ac.uk). The Je-S helpdesk will then advise you of the process and send the appropriate documentation to be completed.

INTERNATIONAL RESEARCHERS

RCUK encourages international collaboration by 'supporting enabling activities and reducing barriers'. This includes the following activities:

- helping researchers overcome the problem of 'double jeopardy' (the risk that a proposed joint project will be approved in one country but not in another);

- establishing partnership links between research institutions;

- building on existing links between research groups and extending networks;

- encouraging researchers from overseas to undertake research in the UK as well as UK researchers to spend time abroad.

RCUK has developed a range of funding programmes to achieve these aims. Many of these schemes are available for early-career researchers. For a current list of calls, visit the RCUK website: www. rcuk.ac.uk/international/funding/FundingOpps.

International offices have been set up in China, India, the US and Europe to aid collaboration. More information and advice can be obtained from the relevant office:

RCUK China
Tel: +86 10 5192 4000
Email: info@rcuk.cn

RCUK India
Tel: +91 11 2419 2370
Email: rcuk.india@rcuk.ac.uk

RCUK US
Tel: +1 202 588 7693
Email: Stephen.Elsby@rcuk.ac.uk

The UK Research Office in Europe
Tel: 00 32 2 230 1535/5275
Email: ukro@bbsrc.ac.uk

Funding Databases (Online)

UNITED KINGDOM
Charity Choice

Charity Choice (www.charitychoice.co.uk) provides details of over 160,000 charities in the UK. You can search the database by charity name or number, by charity sector or by region. The site has been set up to provide a quick and simple way to donate to charity, but it is also a useful resource to help you find out about charities that may be able to provide funds for your research.

Euraxess Funding Search

Euraxess Funding Search (euraxessfunds.britishcouncil.org) is provided by the British Council and is a searchable database that enables researchers to obtain funding for international travel, short research visits and overseas fellowships (for researchers coming from and to the UK).

Funding Central

Funding Central (www.fundingcentral.org.uk) is a free website for charities, voluntary organizations and social enterprises in the UK, providing access to thousands of funding and finance opportunities for voluntary and community organizations operating in England. It is managed by the National Council for Voluntary Organizations in partnership with Idox Information Solutions Ltd and is funded by the Office for Civil Society.

Funder Finder

Funder Finder (www.biglotteryfund.org.uk/funding/funding-finder) is a searchable database provided by the Big Lottery Fund in the UK. It enables community groups, charities, public organizations and not-for-profit groups to identify the most suitable funding programme for their project. Although academic research is no longer funded by the Big Lottery, it might be possible to obtain funds for a community project that involves an element of research.

GRANTfinder 4 education

GRANTfinder 4 education (www.idoxgrantfinder.co.uk/education) is a service provided by Idox Information Solutions Ltd. The database provides information about over 9,000 funding schemes including grants, loans and awards from local, regional and national UK government, European funding, charitable trusts and corporate sponsors. You will need to pay a subscription fee to use the service.

Grant Tracker

Grant Tracker (www.grant-tracker.org) provides details of almost 1,000 funding schemes for charities, clubs, community groups and other not-for-profits in Northern Ireland. You will need to subscribe to use the service. In 2014 it costs £145 for one year, £40 for one month and £10 for one day.

Research Professional

Research Professional (www.researchprofessional.com) provides information about research-based funding opportunities and news about research policy in the UK. You can use the public website (www.researchresearch.com) to find out more about the organization, but you will need to subscribe to access information about specific funding opportunities (most universities subscribe to the service).

The Directory of Social Change

The Directory of Social Change runs four funding sites (www.trustfunding.org.uk, www.governmentfunding.org.uk, www.companygiving.org.uk and www.grantsforindividuals.org.uk). On these sites you can access information about trusts, foundations, government funding, private funding and charitable funding that may help you to pay for your research. However, you have to subscribe and pay a fee to use these services. In 2014 the fees are £285 + VAT (£342) for one user licence for voluntary groups and £466 + VAT (£559.20) for one user licence for commercial/statutory groups. If you are thinking about subscribing, find out first whether your employer/university already has a subscription to any of these sites.

EUROPEAN UNION
Scholarship Portal

The Scholarship Portal (www.scholarshipportal.eu) provides information about scholarships and grants to study and undertake research in Europe. It is a free-to-use search facility that covers undergraduate, postgraduate and staff level study and research.

REPUBLIC OF IRELAND
The Wheel

The Wheel (www.wheel.ie) provides a forum and resources for the voluntary and community sector in Ireland. It includes access to Funding Point, which is a database of around 840 funding organizations. You will need to subscribe to the service. In 2014 it costs members of The Wheel €125 per year, other community and voluntary organizations €200 per year, and all others €400 per year.

UNITED STATES OF AMERICA
Foundation Center

The Foundation Center (foundationcenter.org) provides information about philanthropy worldwide. This includes information about private foundations, community foundations, grant-making public charities and corporate giving programmes. You can use a free online search facility to obtain details of over 90,000 funding organizations.

Grants.gov

Grants.gov (www.grants.gov) enables you to find and apply for federal grants in the US. The database is free to use and you can browse by keyword, newest opportunities, category, agency or eligibility. The website contains some useful information about federal funding and eligibility criteria.

National Science Federation

The National Science Federation (www.nsf.gov) funds research and education in most fields of science and engineering. There is a free-to-use funding search database on the website that enables you to search for funding by keyword, alphabetically or by specialist programme. There is also an advanced search facility available.

AUSTRALIA
The Australian Competitive Grants Register

The Australian Competitive Grants Register (education.gov.au/Australian-competitive-grants-register) lists schemes that provide competitive research grants to higher education providers in Australia. The Register lists funding organizations and specific schemes that are funded by each organization. It is updated every year.

The Australian Directory of Philanthropy

The Australian Directory of Philanthropy (www.philanthropy.org.au) is an online database of grant-making organizations in Australia. The database is useful for grant-seekers, researchers, journalists and not-for-profit groups and is updated annually. An annual subscription costs $99.00 in 2014.

WORLDWIDE
Pivot

Pivot (pivot.cos.com) is aimed at research administrators, research development professionals and their institutions. It combines a comprehensive database of funding opportunities with a database of three million pre-populated scholar profiles, enabling a matching of researcher profiles with funding opportunities. A thirty-day free trial for institutions is available on the website.

The Directory of Development Organizations

The Directory of Development Organizations (www.devdir.org) is a free-to-use service that lists 70,000 development organizations worldwide. The directory is divided into world regions and you can download the directory for each country within this region. Links to websites are available, along with contact details of each organization. You can also create your own personal directory of development organizations, using this website. Although not specifically a funding database, it is a useful resource for finding out about possible sources of funding.

The International Cancer Research Partnership database

The International Cancer Research Partnership (ICRP) database (www.icrpartnership.org/database.cfm) is a free-to-use database of research awards from all member organizations. It is structured in an internationally recognized classification system known as the Common Scientific Outline (CSO). The database enables you to understand what type of funding is available, identify potential collaborators, avoid duplication of effort and find researchers to assist with peer review of grant applications and journal articles.

APPENDIX 3

Funding Directories (Printed)

Lernelius-Tonks, L., *The Guide to Educational Grants 2013/14* (12th edition, London: Directory of Social Change, 2013).

Palgrave Macmillan Ltd (ed), *The Grants Register 2014: The Complete Guide to Postgraduate Funding Worldwide* (32nd edition, Basingstoke: Palgrave MacMillan, 2013).

Traynor, T., *The Directory of Grant Making Trusts 2012–2013* (22nd edition, London: Directory of Social Change, 2012).

Wilmington Publishing & Information Ltd., *Charity Choice UK, Charities Digest, Top 3000 Charities, Charity Choice Scotland, Charity Choice Northern Ireland* (all published annually, London: Wilmington Publishing & Information Ltd).

Wooller, J. and M. Wooller (eds), *The Finance and Funding Directory 2012/13: A Comprehensive Guide to the Best Sources of Finance and Funding* (3rd edition, Petersfield: Harriman House Ltd, 2012).

Further Reading and Resources

CHAPTER 1: KNOWING ABOUT SOURCES OF FUNDING
Further reading

Botting Herbst, N. and Norton, M., *The Complete Fundraising Handbook* (6th edition, London: Directory of Social Change, 2012)

CHAPTER 2: CARRYING OUT A FUNDING SEARCH
Directories

Lernelius-Tonks, L., *The Guide to Educational Grants 2013/14* (12th edition, London: Directory of Social Change, 2013)

Traynor, T., *The Directory of Grant Making Trusts 2012–2013* (22nd edition, London: Directory of Social Change, 2012)

Wooller, J. and Wooller, M. (eds), *The Finance and Funding Directory 2012/13: A Comprehensive Guide to the Best Sources of Finance and Funding* (3rd edition, Petersfield: Harriman House Ltd, 2012)

Useful websites

- Charity Choice www.charitychoice.co.uk
- Euraxess Funding Search euraxessfunds.britishcouncil.org
- Foundation Center foundationcenter.org
- Funder Finder www.biglotteryfund.org.uk/funding/funding-finder

- GRANTfinder 4 education www.idoxgrantfinder.co.uk/ education

- Grant Tracker www.grant-tracker.org

- Research Professional www.researchprofessional.com

- The Wheel www.wheel.ie

- Trustfunding www.trustfunding.org.uk

- For more information see Appendix 2.

CHAPTER 3: FUNDING FOR ACADEMIC RESEARCHERS
Further reading

Berry, D., *Gaining Funding for Research: A Guide for Academics and Institutions* (Maidenhead: Open University Press, 2010)

Lawton, K. and Marom, D., *The Crowdfunding Revolution: How to Raise Venture Capital Using Social Media* (New York, NY: McGraw-Hill Professional, 2013)

Directories

Lernelius-Tonks, L., *The Guide to Educational Grants 2013/14* (12th edition, London: Directory of Social Change, 2013)

Palgrave Macmillan Ltd (ed), *The Grants Register 2014: The Complete Guide to Postgraduate Funding Worldwide* (32nd edition, Basingstoke: Palgrave MacMillan, 2013)

Useful websites

- Euraxess Funding Search euraxessfunds.britishcouncil.org

- Foundation Center foundationcenter.org

- GRANTfinder 4 education www.idoxgrantfinder.co.uk/ education

- National Science Federation www.nsf.gov

- Research Professional www.researchprofessional.com

- Trustfunding www.trustfunding.org.uk

For more information see Appendix 2.

The research council websites in the UK are as follows:

- Arts and Humanities Research Council www.ahrc.ac.uk
- Biotechnology and Biological Sciences Research Council www.bbsrc.ac.uk
- Engineering and Physical Sciences Research Council www.epsrc.ac.uk
- Economic and Social Research Council www.esrc.ac.uk
- Medical Research Council www.mrc.ac.uk
- Natural Environment Research Council www.nerc.ac.uk
- Science and Technology Facilities Council www.stfc.ac.uk

CHAPTER 4: FUNDING FOR PRIVATE SECTOR AND INDUSTRY RESEARCHERS
Directories

Wooller, J. and Wooller, M. (eds), *The Finance and Funding Directory 2012/13: A Comprehensive Guide to the Best Sources of Finance and Funding* (3rd edition, Petersfield: Harriman House Ltd., 2012)

Useful websites

- The Directory of Social Change www.companygiving.org.uk

For more information see Appendix 2

CHAPTER 5: FUNDING FOR CHARITY, COMMUNITY AND NOT-FOR-PROFIT RESEARCHERS
Further reading

Whaley, S., *Fundraising for a Community Project: How to Research Grants and Secure Financing for Local Groups and Projects in the UK* (Oxford: How to Books, 2007)

Directories

Traynor, T., *The Directory of Grant Making Trusts 2012–2013* (22nd edition, London: Directory of Social Change, 2012)

Wilmington Publishing & Information Ltd, *Charity Choice UK, Charities Digest, Top 3000 Charities, Charity Choice Scotland, Charity Choice Northern Ireland* (all published annually, London: Wilmington Publishing & Information Ltd).

Useful websites

- Charity Choice www.charitychoice.co.uk

- Funding Central www.fundingcentral.org.uk

- Funder Finder www.biglotteryfund.org.uk/funding/funding-finder

- Grant Tracker www.grant-tracker.org

- The Directory of Development Organizations www.devdir.org

- The Wheel www.wheel.ie

- Trustfunding www.trustfunding.org.uk

For more information see Appendix 2.

CHAPTER 6: FUNDING FOR SELF-EMPLOYED AND RETIRED RESEARCHERS
Further reading

Collett, P. and Fenton, W., *The Sponsorship Handbook: Essential Tools, Tip and Techniques for Sponsors and Sponsorship Seekers* (San Francisco, CA: Jossey-Bass, 2011)

Lawton, K. and Marom, D., *The Crowdfunding Revolution: How to Raise Venture Capital Using Social Media* (New York, NY: McGraw-Hill Professional, 2013)

Directories

Wooller, J. and Wooller, M. (eds), *The Finance and Funding Directory 2012/13: A Comprehensive Guide to the Best Sources of Finance and Funding* (3rd edition, Petersfield: Harriman House Ltd., 2012)

Useful websites

- Charity Choice www.charitychoice.co.uk

- Funding Central www.fundingcentral.org.uk

- The Directory of Development Organizations www.devdir.org

- The Directory of Social Change www.grantsforindividuals.org.uk

- Trustfunding www.trustfunding.org.uk

For more information see Appendix 2.

CHAPTER 7: CHOOSING A FUNDING ORGANIZATION
Useful websites

The Bill and Melinda Gates Foundation (www.gatesfoundation.org) provides helpful information about how the organization works and how they make grants. It is a good example of the type of information that you should seek to determine whether you can ensure compatibility of subject, purpose and ethos.

CHAPTER 8: KNOWING ABOUT COSTING METHODS
Useful websites

Research Councils UK (www.rcuk.ac.uk) has some helpful documents available for download from their website. For example, you might find 'Full Economic Costing: Updated guidance notes for peer reviewers' useful as it explains very clearly how fEC should be considered throughout the peer reviewer process.

CHAPTER 9: WORKING OUT COSTS
Useful websites

HEFCE (www.hefce.ac.uk) provides some helpful information about research funding and costing for research in the UK. This includes information about the UK Research Partnership Investment Fund (set up in 2012 to support investment in higher education research facilities), mainstream quality-related research funding (distributed on the basis of research quality) and capacity funding (supporting research capacity development in science, engineering and technology).

Jisc (www.jisc.ac.uk/advice/reducing-costs) provides helpful information about how you can reduce ICT costs in education and research, which will help you to demonstrate value for money in your grant application.

CHAPTER 10: WORKING WITH COLLABORATORS
Useful websites

The Directory of Social Change has a website (www.companygiving. org.uk) that helps you to research potential corporate partnerships. There is a fee to pay if you wish to subscribe to this service.

The National Institute for Health Research has a useful webpage for researchers in the UK who are thinking about collaborating with industry: www.nihr.ac.uk/industry.

The Academy of Medical Sciences is keen to 'encourage permeability' between universities, industry and the NHS in the UK. More information about initiatives that have been set up to achieve this can be obtained from their website: acmedsci.ac.uk.

Knowledge Transfer Partnerships (KTPs) support 'UK businesses wanting to improve their competitiveness, productivity and performance by accessing the knowledge and expertise available within

UK Universities and Colleges'. More information about KTPs can be obtained from their website: www.ktponline.org.uk.

CHAPTER 11: JUSTIFYING YOUR BUDGET
Useful websites

HEFCE (www.hefce.ac.uk) provides some helpful information about obtaining value for money in research. You can also access the Financial Memorandum between HE institutions and HEFCE and the HEFCE Audit Code of Practice on this website.

CHAPTER 12: FINALIZING YOUR BUDGET
Useful websites

Visit the Australian Taxation Office website (www.ato.gov.au) for more information about the Goods and Services Tax (GST).

CHAPTER 13: SEEKING GRANT APPLICATION ADVICE
Further reading

Browning, B., *Perfect Phrases for Writing Grant Proposals: Hundreds of Ready-to-Use Phrases to Present Your Organization, Explain Your Cause and get the Funding you Need* (New York, NY: McGraw-Hill Professional, 2008)

O'Neal-McElrath, T., *Winning Grants Step-by-Step: The Complete Workbook for Planning, Developing and Writing Successful Proposals* (4th edition, San Francisco, CA: Jossey-Bass, 2013)

CHAPTER 14: PLANNING YOUR GRANT APPLICATION
Further reading

Jansen, R. C., *Funding your Career in Science: From Research Idea to Personal Grant* (New York: Cambridge University Press, 2013)

Payne, M. A., *Grant Writing Demystified* (New York, NY: McGraw-Hill Contemporary, 2011)

Sternberg, R. J., (ed) *Writing Successful Grant Proposals from the Top Down and Bottom Up* (Thousand Oaks, CA: Sage Publications Inc., 2014)

CHAPTER 15: WRITING YOUR GRANT APPLICATION
Further reading

Denicolo, P. (ed), *Achieving Impact in Research* (London: Sage, 2013)

Denicolo, P. and Becker, L., *Developing Research Proposals* (London: Sage, 2012)

Denscombe, M., *Research Proposals: A Practical Guide* (Maidenhead: Open University Press, 2012)

Pryor, G. (ed), *Managing Research Data* (London: Facet Publishing, 2012)

Punch, K., *Developing Effective Research Proposals* (2nd edition, London: Sage, 2006).

White, P., *Developing Research Questions: A Guide for Social Scientists* (London: Palgrave Macmillan, 2009)

Useful websites

The Digital Curation Centre (www.dcc.ac.uk) is a 'centre of expertise in digital information curation with a focus on building capacity, capability and skills for research data management across the UK's higher education research community'. Useful information about data management is available on this website.

CHAPTER 16: SUBMITTING YOUR GRANT APPLICATION
Useful websites

Research Councils UK (www.rcuk.ac.uk) has some useful information about making and submitting grant applications to research councils in the UK. This includes helpful submission guidance notes and comprehensive information about the peer review process.

CHAPTER 17: MAKING SUCCESSFUL APPLICATIONS
Further reading

Aldridge, J. and Derrington, A., *The Research Funding Toolkit: How to Plan and Write Successful Grant Applications* (London: Sage, 2012)

Eastwood, M. and Norton, M., *Writing Better Fundraising Applications: A Practical Guide* (4th edition, London: Directory of Social Change, 2010)

Warwick, M., *How to Write Successful Fundraising Letters* (2nd edition, San Francisco, CA: Jossey-Bass, 2008)

CHAPTER 18: COPING WITH UNSUCCESSFUL APPLICATIONS
Further reading

Collins, V., *50 Mistakes Grant Writers Make: Easy to Use Tips for Writing Winning Grant Proposals* (Floyds Knobs, IN: Heart Thoughts Publishing, 2013)

Further information

Increasingly, researchers are writing blogs about grant rejection. These provide useful and interesting personal information about what it feels like to be unsuccessful in your grant application and what you should do when your proposal has been rejected. The search term 'handling grant rejection' takes you to some of the more interesting blogs.

CHAPTER 19: ADDRESSING ETHICAL ISSUES
Further reading

Comstock, G., *Research Ethics: A Philosophical Guide to the Responsible Conduct of Research* (New York: Cambridge University Press, 2013)

Ransome, P., *Ethics and Values in Social Research* (London: Palgrave Macmillan, 2013)

Wiles, R., *What are Qualitative Research Ethics?* (London: Bloomsbury Academic, 2013)

Useful documents

The Economic and Social Research Council (ESRC) has a document called the *ESRC Framework for Research Ethics*, which lays out the principles, procedures and minimum requirements for all types of research supported by the ESRC. A copy of the document can be downloaded from the ESRC website: www.esrc.ac.uk.

The British Sociological Association has a *Statement of Ethical Practice* that can be downloaded in full from their website: www. britsoc.co.uk. This is a comprehensive statement that covers issues such as professional integrity, relationships with research participants, anonymity, confidentiality, privacy and obligations to sponsors and funders.

You can download a publication called *Ethical Guidelines*, which has been produced by the Social Research Association: the-sra.org. uk. This is a full and comprehensive guide that covers all ethical issues for social researchers.

CHAPTER 20: ADDRESSING CONFLICT OF INTEREST
Further reading

Hammersley, M. and Traianou, A., *Ethics in Qualitative Research: Controversies and Contexts* (London: Sage, 2012)

CHAPTER 21: AVOIDING BIASED FINANCIAL RELATIONSHIPS
Further reading

Miller, T., Birch, M., Mauthner, M. and Jessop, J., (eds), *Ethics in Qualitative Research* (2nd edition, London: Sage, 2012)

CHAPTER 22: APPLYING FOR ETHICAL APPROVAL
Useful booklets and tools

The National Research Ethics Service (www.nres.nhs.uk) has produced a booklet called *Defining Research* for researchers in the UK. It can be downloaded from the NRES website.

The Medical Research Council has produced a Health Research Authority Decision Tool (www.hra-decisiontools.org.uk/research) that helps you to find out whether your work is classed as research.

CHAPTER 23: USING ETHICAL APPROVAL AND REVIEW BODIES
Websites and email addresses

Australian Health Ethics Committee and Human Research Ethics Committees (AUS)

Website: www.nhmrc.gov.au/health-ethics/human-research-ethics

Email: ethics@nhmrc.gov.au

European Network of Research Ethics Committees (EU)

Website: www.eurecnet.org

Email: lanzerath@eurecnet.org

Office for Human Research Protections (US)

Website: www.hhs.gov/ohrp

Email: OHRP@hhs.gov

IRB database: ohrp.cit.nih.gov/search

The Health Research Authority (UK)

Website: www.hra.nhs.uk

Email: contact.hra@nhs.net

The National Council on Ethics in Human Research (CAN)

Website: www.ncehr-cnerh.org

The National Ethics Advisory Committee (NZ)

Website: neac.health.govt.nz

Email: neac@moh.govt.nz

Directories

You can obtain contact details of all the NHS Research Ethics Committees in the UK by visiting www.hra.nhs.uk/resources. You can subscribe to the REC Directory RSS feed to keep up-to-date with changes to REC contact details.

Bibliography

Aldridge, J. and Derrington, A., *The Research Funding Toolkit: How to Plan and Write Successful Grant Applications* (London: Sage, 2012)

Berry, D., *Gaining Funding for Research: A Guide for Academics and Institutions* (Maidenhead: Open University Press, 2010)

Botting Herbst, N. and Norton, M., *The Complete Fundraising Handbook* (6th edition, London: Directory of Social Change, 2012)

Browning, B., *Perfect Phrases for Writing Grant Proposals: Hundreds of Ready-to-Use Phrases to Present Your Organization, Explain Your Cause and get the Funding You Need* (New York, NY: McGraw-Hill Professional, 2008)

Collett, P. and Fenton, W., *The Sponsorship Handbook: Essential Tools, Tip and Techniques for Sponsors and Sponsorship Seekers* (San Francisco, CA: Jossey-Bass, 2011)

Denicolo, P. and Becker, L., *Developing Research Proposals* (London: Sage, 2012)

Denscombe, M., *Research Proposals: A Practical Guide* (Maidenhead: Open University Press, 2012)

Eastwood, M. and Norton, M., *Writing Better Fundraising Applications: A Practical Guide* (4th edition, London: Directory of Social Change, 2010)

Hall, J. L., *Grant Management: Funding for Public and Non-Profit Programs* (Sudbury, MA: Jones and Bartlett Inc., 2010)

Hammersley, M. and Traianou, A., *Ethics in Qualitative Research: Controversies and Contexts* (London: Sage, 2012)

Lawton, K. and Marom, D., *The Crowdfunding Revolution: How to Raise Venture Capital Using Social Media* (New York, NY: McGraw-Hill Professional, 2013)

Miller, T., Birch, M., Mauthner, M. and Jessop, J. (eds), *Ethics in Qualitative Research* (2nd edition, London: Sage, 2012)

O'Neal-McElrath, T., *Winning Grants Step-by-Step: The Complete Workbook for Planning, Developing and Writing Successful Proposals* (4th edition, San Francisco, CA: Jossey-Bass, 2013)

Payne, M. A., *Grant Writing Demystified* (New York, NY: McGraw-Hill Contemporary, 2011)

Punch, K., *Developing Effective Research Proposals* (2nd edition, London: Sage, 2006)

Ransome, P., Ethics and Values in Social Research (London: Palgrave Macmillan, 2013)

Sternberg, R. J., (ed) *Writing Successful Grant Proposals from the Top Down and Bottom Up* (Thousand Oaks, CA, Sage Publications Inc., 2014)

Warwick, M., *How to Write Successful Fundraising Letters* (2nd edition, San Francisco, CA: Jossey-Bass, 2008)

Whaley, S., *Fundraising for a Community Project: How to Research Grants and Secure Financing for Local Groups and Projects in the UK* (Oxford: How to Books, 2007)

White, P., *Developing Research Questions: A Guide for Social Scientists* (London: Palgrave Macmillan, 2009)

Wiles, R., *What are Qualitative Research Ethics?* (London: Bloomsbury Academic, 2013).

Index